Beyond Close Air Support

Forging a New Air-Ground Partnership

Bruce R. Pirnie, Alan Vick, Adam Grissom,
Karl P. Mueller, David T. Orletsky

Prepared for the United States Air Force
Approved for public release; distribution unlimited

The research described in this report was sponsored by the United States Air Force under Contract F49642-01-C-0003. Further information may be obtained from the Strategic Planning Division, Directorate of Plans, Hq USAF.

Library of Congress Cataloging-in-Publication Data

Beyond close air support : forging a new air-ground partnership / Bruce R. Pirnie ... [et al.].
p. cm.
Includes bibliographical references.
"MG-301."
ISBN 0-8330-3741-2 (pbk.)
1. Close air support. 2. Military doctrine—United States. 3. United States. Air Force. I. Pirnie, Bruce, 1940–

UG700.B48 2005
358.4'142—dc22

2004030608

Published 2005 by the RAND Corporation
1776 Main Street, P.O. Box 2138, Santa Monica, CA 90407-2138
1200 South Hayes Street, Arlington, VA 22202-5050
201 North Craig Street, Suite 202, Pittsburgh, PA 15213-1516
RAND URL: http://www.rand.org/
To order RAND documents or to obtain additional information, contact
Distribution Services: Telephone: (310) 451-7002;
Fax: (310) 451-6915; Email: order@rand.org

Preface

Although airmen have been providing close air support (CAS) to friendly ground forces since World War I, recent operations in Afghanistan and Iraq have brought renewed attention to the unique demands of this mission. The Army increasingly views air power as indispensable to its future warfighting concepts and seeks mechanisms to ensure that it is available and responsive to the needs of the land forces. For the Air Force, counterland operations are becoming more important, but airmen remain concerned with ensuring that air power's unique ability to mass rapidly is not lost in efforts to provide on-call fires to small ground elements spread across a large battle space.

To address these and related policy challenges, Project AIR FORCE conducted a study of close support on the future battlefield. The study addressed three major policy questions: (1) How should air attack and ground maneuver be integrated? (2) How should the CAS terminal attack control function be executed? (3) How should ground maneuver/fires and air attack be deconflicted? This research builds on work done in Project AIR FORCE over the past ten years to provide a better understanding of the air-ground partnership as well as to enhance the Air Force contribution in operations against enemy land forces. Previous RAND reports in this area include:

- *The Stryker Brigade Combat Team: Rethinking Strategic Responsiveness and Assessing Deployment Options*, by Alan Vick, David T. Orletsky, Bruce Pirnie, and Seth G. Jones, MR-1606-AF, 2002.

- *Aerospace Operations Against Elusive Ground Targets*, by Alan Vick, Richard M. Moore, Bruce Pirnie, and John Stillion, MR-1398-AF, 2001.
- *Aerospace Operations in Urban Environments: Exploring New Concepts*, by Alan Vick, John Stillion, Dave Frelinger, Joel S. Kvitky, Benjamin S. Lambeth, Jefferson P. Marquis, and Matthew C. Waxman, MR-1187-AF, 2000.
- *Enhancing Airpower's Contribution Against Light Infantry Targets*, by Alan Vick, John Bordeaux, David T. Orletsky, and David A. Shlapak, MR-697-AF, 1996.

The research reported here was sponsored by the Director of Operational Planning, Headquarters, U.S. Air Force, and was conducted within the Strategy and Doctrine Program of RAND Project AIR FORCE.

RAND Project Air Force

RAND Project AIR FORCE (PAF), a division of the RAND Corporation, is the U.S. Air Force's federally funded research and development center for studies and analyses. PAF provides the Air Force with independent analyses of policy alternatives affecting the development, employment, combat readiness, and support of current and future aerospace forces. Research is conducted in four programs: Aerospace Force Development; Manpower, Personnel, and Training; Resource Management; and Strategy and Doctrine. The research reported here was prepared under contract F49642-01-C-0003.

Additional information about PAF is available at http://www.rand.org/paf.

Contents

Figures

Tables

Summary

Recent operations in Afghanistan and Iraq have reawakened interest in counterland operations. One battle in particular, Operation Anaconda in Afghanistan, sparked a heated debate between the Air Force and the Army about the conduct of close air support (CAS) and led to new efforts to improve the integration of air power and ground power prior to Operation Iraqi Freedom. Although these efforts were quite successful, there is growing recognition by both airmen and soldiers that air-ground cooperation is increasingly important and that additional steps must be taken.

This report seeks to help the Air Force engage the Army in a constructive dialogue on this issue. In particular, it addresses three policy questions: (1) How should air attack and ground maneuver be integrated? (2) How should the CAS terminal control function be executed? (3) How should ground maneuver/fires and air attack be deconflicted?

The Evolving Relationship Between Air Power and Land Power (see pp. 20–30)

Whether air power or land power should predominate depends on the particular military problem being considered. Depending upon the situation, either might predominate, and their relationship is likely to shift over the course of a campaign. At one extreme, air power might augment the firepower of ground units, even replacing artillery in some cases. Marines habitually take this approach, and it might also be valid for Army forces in some situations, such as an airborne assault. At the other extreme, air power might coerce an opponent or destroy his military forces in the absence of any ground operation.

Between the extremes are three plausible alternatives, one highlighting air power, one highlighting land power, and one based on partnership. From the perspective of a land-force commander, air power's greatest contribution is in weakening and impeding enemy forces before they can close with friendly troops. From the perspective of an air-force commander, land power's greatest contribution is in flushing and fixing enemy forces so that they can be destroyed by air attack. Both perspectives are valid, but neither captures the whole truth. The most fruitful perspective is a partnership in which either or neither partner may predominate, depending upon the operational and tactical situation.

There are several reasons for developing a partnership. It is the approach most suitable to the largest number of adversaries, and it can easily be adjusted toward greater prominence for either partner. It gives the least opportunity for parochial claims and one-sided pleading for one's own service. Its very difficulty could be a virtue: Once the services have mastered partnership, they can easily revert to simpler approaches.

Partnership does not, of course, imply having co-equal commanders of the same operation, thus violating unity of command. It implies an allocation of authority that maximizes the contributions of each partner toward a common endeavor. Within the range of his organic weapons (normally 30 to 40 kilometers), the land-force commander rightly expects to control air attacks. Indeed, he must have such control in order to integrate direct fires, artillery, rockets, attack helicopters, and fixed-wing aviation. Beyond that range, an air-force commander should control air attacks, but with a view to assuring successful maneuver of land forces. Neither of these commanders need be considered supported or supporting according to doctrine, since both work for the same joint-force commander.

Trends in Counterland Operations (see pp. 31–38)

Enemy land forces were the critical target set during recent conflicts in Kosovo, Afghanistan, and Iraq. In all of these conflicts, enemy land forces were the only target set that was undeniably legitimate, politically acceptable, and of pivotal importance. Destroying Serb ground forces in Kosovo would have been the most direct way to accomplish NATO's goal of ending the oppression of Kosovar Albanians by Serbia. Attacking Taliban ground forces in Afghanistan toppled the Taliban regime, stripping al Qaeda of its sanctuary. Defeating Iraqi ground forces assured the end of Ba'athist rule and made Saddam Hussein a hunted fugitive. Moreover, in all three cases, there were cogent political reasons for avoiding extensive damage to infrastructure. The case of Kosovo is particularly instructive because when Milosevic capitulated, NATO had almost exhausted the targets its members were willing to strike.

When accomplished jointly, counterland operations by air forces are becoming ever more effective. Thanks to improved sensors and precision munitions, air attacks are now effective at night, during extreme weather conditions, and in close proximity to friendly forces. Moreover, the potential for fratricide is declining, thanks to improved communications and tracking of friendly ground forces through the Global Positioning System (GPS). The chief impediment to successful counterland operations is the inability to detect and identify enemy ground forces. Again, the Kosovo case is particularly instructive. In the absence of a credible threat from NATO land forces, Serb forces were free to disperse and hide in terrain that offered plenty of cover and concealment. As a result, air attacks against them were not effective. Indeed, Serb forces drove hundreds of thousands of Kosovar civilians from their homes during the NATO bombing at little cost to themselves. In contrast, indigenous opposition forces fixed Taliban forces, making them easy targets for air attack, and coalition land forces flushed Iraqi forces, making them reveal their positions.

Jointness is descending to lower levels of command, but current doctrine was designed for the Cold War, when jointness tended to reside at higher levels. Special operations forces (SOF), employed more frequently in recent years, take jointness down to the level of very small teams. An Operational Detachment-Alpha in the Army's Special Forces is just a squad, yet it may operate independently, and, normally augmented with terminal attack controllers (TACs), it may call in large numbers of air attacks. Conventional forces are operating at lower force levels, implying that jointness has to descend to lower levels. In Afghanistan, for example, U.S. land forces consisted of just one understrength brigade at the height of combat operations. In the combat phase of operations in Iraq, the Army fielded a corps headquarters and two full divisions, but only one division led the advance, and it usually had one brigade in front. The tendency to push jointness down to lower levels will probably accelerate as the Army fields new forces that operate in more fluid fashion.

Key Findings (see pp. 167–170)

Key findings of this study are summarized below.

- **Army Transformation is increasing Army interest in air attack.** As the Army seeks to become more strategically deployable and agile on the battlefield, it is reducing the weight of ground-based fires available to maneuver units. Although not yet fully detailed, the number of independent artillery brigades will shrink as the Army shifts manpower in those units to military police and other undermanned functions. Moreover, operations are expected to center increasingly on independent brigades, which will operate without or with less division and corps fire support. These factors, combined with a newfound Army confidence in the accuracy and responsiveness of air-delivered fires, will result in increased Army requests for CAS and air interdiction.
- **Army Transformation will increase the demand for terminal attack controllers.** Current joint procedures require that a certified TAC control aircraft conducting normal CAS missions. The Army wants to have this capability at company level. To satisfy

this demand, the Air Force must either train more TACs or change the way they are organized.

- **The joint terminal attack controller (JTAC) program is not designed to generate a large number of certified TACs.** The JTAC program was created to ensure that TAC standards are uniform across the services, not to produce a vast new pool of TACs. Whether TACs are trained at a joint school or produced by the services, the fundamental constraints remain the same: a shortage of qualified candidates, a demanding job that takes years to master, a shortage of training facilities (ranges and simulators), and heavy demands on strike aircraft that make it difficult for them to generate the necessary training sorties for more than the current TAC force.
- **Operational/technological trends and manpower realities, not service preferences, are at the heart of the TAC debate.** Some view the TAC debate as the latest event in a long struggle between airmen and soldiers over the control of air power. In our judgment, however, the debate is driven by operational and manpower realities, not service preferences or doctrine. The Army recognizes a strong trend toward dispersion on the battlefield and is appropriately adapting its forces to operate in smaller elements dispersed across a larger battle space. Such forces will need more ready and routine access to air power. The Air Force is correct in insisting that only fully certified, experienced, and proficient TACs have the authority to control aircraft.
- **Creative use of available technologies can free TACs to focus on essential functions and can give engaged ground elements greater access to joint fires.** The Army does not really need TACs with every engaged combat unit. What it needs is a system that allows engaged elements to designate targets, TACs, and fire support officers (FSOs) at the battalion level to confirm that no friendly forces are at the target locations, and aircrews to independently confirm that the targets are good. The technologies discussed in Chapter Six would enable such a system. These technologies already exist or are well along in development.

- **Disaggregating the TAC function is essential to ensuring that both Army and Air Force battlefield needs are met.** Identifying TAC functions that could be delegated to engaged combat units (e.g., target identification and geolocation) would ensure that dispersed ground elements could easily call for air support and would allow TACs to focus on those functions that require a fully certified controller (e.g., aircraft control and deconfliction). It is the only option that has a high probability of meeting Army needs without presenting undue risk to ground and air forces.
- **The doctrine for counterland operations and the associated control measures needs revision.** In current counterland doctrine, only CAS is satisfactorily defined; interdiction is poorly defined; and strategic attack is barely mentioned. These missions should be redefined with greater clarity, linking them unambiguously to the actual and contemplated actions of maneuver forces. In current doctrine, the fire support coordination line (FSCL) is unrelated to missions and is often contentious. During operations in Iraq, the 3rd Infantry Division almost overran the FSCL because the FSCL could not be adjusted quickly enough. At other times, the line was placed too far ahead, imposing unnecessary and counterproductive constraints on air attack. It should be redefined as the outer edge of CAS, usually at about artillery range beyond friendly forces, i.e., the area where integration of fires is necessary. As command and control matures, the FSCL should be replaced with a flexible system of kill boxes. The CAS area would be defined as those kill boxes where terminal attack control, implying control by a land-force commander, is required. Outside this area, an air-force commander should have the authority to conduct the counterland mission, always assuming that his efforts will complement and not run counter to the scheme of maneuver.
- **Army organic fires remain the most efficient means of meeting routine unplanned requests.** Army standards for responsiveness in counterbattery fire are high. For example, counterbattery fire was often delivered within two minutes of sensing the incoming fire during Operation Iraqi Freedom. This level of responsive-

ness is possible from the air for selected high-priority missions (e.g., the leading elements in a major offensive such as the 3/7th Cavalry during Operation Iraqi Freedom, or Special Forces conducting direct-action missions) but requires a huge force structure to sustain for prolonged operations over a large battle space. New concepts for long-range joint fires might meet some of these needs, but the most responsive systems (missiles) tend to be extremely costly and are often inappropriate for small-unit needs—and even missiles cannot meet single-digit response times unless they are relatively close or have hypersonic speed. Therefore, the Army should retain sufficient organic fires to meet the routine fire support needs of dispersed units. Air forces are best used to directly attack enemy maneuver forces throughout the depth of the battlefield, to support selected forces at high risk, to partner with ground forces in planned offensive operations, and to act as a theater reserve.

- **Air attack and ground maneuver should be planned as mutually enabling activities.** "Close air support" is an inaccurate term that implies a one-sided relationship. In modern combat, air and ground forces increasingly operate in mutually enabling ways. This partnership should be encouraged. "Close air attack" is a more accurate description of what modern air forces do in partnership with ground elements. Whenever possible, air elements should be free to conduct deep operations against enemy maneuver forces, thereby isolating the battlefield. These operations have the potential to deny the operational level of maneuver to enemy motorized forces, preventing them from conducting offensive operations at the brigade or higher level. On the isolated battlefield, friendly ground forces can operate in smaller, more dispersed elements, finding and fixing enemy elements that increasingly will operate in small units to minimize their signature. Air and ground forces will attack these forces cooperatively, with air aggressively seeking enemy forces beyond the immediate line of sight of engaged friendly forces and also providing direct support to friendly forces as needed. Finally, in this vision, ground forces do those things they are uniquely able to do: capture and hold territory, find and control weapons of mass destruction (WMD), and enforce peace.

Recommendations for the Air Force and the Army (see pp. 170–171)

As we look to the future, the opportunities for effective partnering of air and ground forces are likely to grow significantly. We recommend that the Army and the Air Force work together to develop new concepts and technologies to speed this process. In particular, training, education, and doctrine will need to be adapted to more smoothly integrate air attack and ground maneuver; the TAC function will need to be disaggregated and new processes developed to effectively designate targets while ensuring that essential oversight remains with the TAC and the combat aircrew; and improved control mechanisms will be needed to exploit the benefits of the digital battlefield and get maximum benefit from the ability of air power to roam over the battlefield.

As adversaries adapt and move away from massed motorized forces operating in the open to dispersed, smaller forces exploiting difficult terrain, a well practiced and developed air-ground partnership will be increasingly necessary.

Acknowledgments

The authors wish to thank the individuals listed below for their assistance to the study. We particularly appreciate the contribution of those officers and NCOs in operational units from all the services who graciously hosted our visits between real-world deployments.

Although this research was conducted for the Air Force, it benefited greatly from interactions with personnel from all the services and other Department of Defense (DoD) organizations. We especially want to acknowledge the insights we gained from related RAND work for the Army. The lead author of this report was also a member of RAND Arroyo Center research teams that conducted studies of Operation Allied Force, Operation Enduring Freedom, and Operation Iraqi Freedom. In the course of research on Operation Iraqi Freedom, he interviewed personnel from V Corps, 3rd Infantry Division (Mechanized), 15th Air Support Operations Squadron (ASOS), 101st Airborne Division, and 2nd Marine Expeditionary Brigade (the core of Task Force Tarawa). General officers interviewed include Lt. Gen. William S. Wallace, Commander V Corps; Maj. Gen. Buford Blount, Commander 3rd Infantry Division (Mechanized); and Brig. Gen. Richard F. Natonski, Commander, 2nd Marine Expeditionary Brigade. Those interviewed at 15th Air Support Operations Squadron include Lt. Col. Mark Bronakowski, Air Liaison Officer (ALO), 3rd Infantry Division; Capt. Jon E. Chesser, ALO, 1st Brigade; Capt. Marco Parzycn, ALO, 1st Brigade; Capt. Charles Glasscock, ALO, 2nd Brigade; SSgt. Travis D. Crosby,

ETAC, Task Force 3-69 Armor, 1st Brigade; and TSgt. Kevin A. Butler, ETAC, 2nd Brigade.

Maj. Michael Pietrucha, the project action officer, provided outstanding support to the study on both substantive and administrative matters. We greatly appreciate his detailed and constructive comments on an earlier draft of this report.

We thank Col. Carl Fosnaugh III (USMC) for inviting us to brief the J-8 Force Application Working Group and Force Application Capabilities Board meetings that he chaired. We thank Lt. Col. Tom Fritz, J-8, who arranged these briefings and took on the huge job of coordinating and integrating comments on our draft report from the Army, USAF, USMC, Navy, JFCOM, CENTCOM, USFK, USAFE, and EUCOM. We thank the anonymous reviewers from these organizations for their careful reviews and suggestions for improvement.

Lt. Col. Steven Kirkpatrick, Commander 93rd Bomb Squadron, and Lt. Col. Blade Walker, Commander 47th Fighter Squadron, supported our visit to the 917th Wing, Barksdale Air Force Base, Louisiana, and made the initial arrangements. Lt. Col. Scott Forrest, 917th Wing, helped organize the visit and acted as escort and tutor on bomber operations. Maj. Jeff Swanson, Air Force Reserve Command Headquarters, helped organize our visit and also gave us a hands-on briefing on-board a B-52. Major Jim "Slick" Travis, 47th Fighter Squadron, provided an illuminating overview of the A-10 and the CAS mission. Other squadron members in the 93rd Bomb Squadron and the 47th Fighter Squadron participated in informative group discussions with us during our visit.

Col. Michael Longoria, the former commander of the 18th Air Support Operations Group, Pope Air Force Base, North Carolina, and now Director of the Joint Air Ground Operations Directorate at Air Combat Command, was an early supporter of this research. Both Col. Longoria and Col. Keith Gentile, his successor, provided exceptional opportunities for our project team to engage the tactical air control party (TACP) community. Lt. Col. Bromwell, ADO of the 18th Air Support Operations Group, and Randall Long arranged interviews and provided access to data during our visits. Lt. Col. John

Masotti, Lt. Col. Franklin Walden, Lt. Col. Pat Pope, Lt. Col. David Hume, Lt. Col. Kermit Phelps, Lt. Col. Seth Bretscher, Capt. Joe Locke, Capt. Jon Shultz, CMSgt. Martin Klukas, Senior MSgt. Chris Griffin, Senior MSgt. Art Boyer, Senior MSgt. Roger Cross, MSgt. William Propst, TSgt. Edward Shulman, TSgt. Rick Winegardner, and Sgt. Stephen Tomat provided comments on our briefing and shared their insights from recent combat operations.

Lt. Col. Byron Risner, former commander of the 15th ASOS, visited RAND to share his insights from Operation Iraqi Freedom.

Gary Vycital at Air Force Special Operations Command Headquarters organized our visit to Hurlburt Field, Florida. Lt. Col. Mike Plehn, commander of the Gunship School, gave us a briefing on experiments in joint air-ground simulations and a tour of the school. At the Tactical Air Control Party School, we met with MSgt. Brett Ramos, TSgt. Mike Brown, SSgt. Charles Keebaugh, and CW3 Ernest Gizoni (US Army SF). The following personnel from AFSOC HQ, the 16th Special Operations Wing, and other AFSOC organizations participated in a roundtable discussion with our team: Lt. Col. Mark Hicks, Lt. Col. Bob Morrow, Lt. Col. Scott Howell, Lt. Col. Jean Paprocki, Maj. Jason Miller, Maj. Edward Espinoza, Capt. Jeff Blackmon, Capt. Wendy Ruffner, CMSgt. Bill Walter, Senior MSgt. Randy Smith, MSgt. Art Ziegler, MSgt. Chris Legg, Paul Brousseau, Geoffrey Hitchcock, Ed McDonald, and Mike Stephens.

At the 720th Special Tactics Group, Hurlburt Field, Maj. Dave Mallamee, Capt. Mike Martin, TSgt. Chris Crutchfield, TSgt. Alan T. Yoshida, William O'Brien, and TSgt. Steve Barrons participated in a roundtable discussion. TSgt. Alan T. Yoshida and Col. Dan Isbell, Air Force Research Laboratory, briefed us on Air Force terminal attack control initiatives. We also thank Col. Craig Rith, then commander, and other personnel in the 720th for their thorough review of and helpful suggestions on our draft report.

At Headquarters Air Combat Command, Lt. Col. Muck Brown shared his CAS expertise on multiple occasions.

At the Marine Corps Combat Development Command, Quantico, Virginia, Col. Art Corbett shared his insights on CAS and fire support.

We thank Maj. Gen. Michael Maples, then commanding general, U.S. Army Field Artillery Center and Ft. Sill, for hosting a project visit to discuss joint fires with him and his senior staff. Sam Coffman organized the visit and provided invaluable data on Army fire support systems.

We thank Maj. Gen. David MacGhee, then commander, Air Force Doctrine Center, for hosting our visit and for including us in an Air Force workshop on counterland doctrine. Col. Tom Ehrhard, Lt. Col John Terino, Harold Winton, and other faculty and students of the School of Advanced Air and Space Studies provided comments on the project briefing and shared their experiences in air-ground operations.

We thank Col. Matt Neuenswander, former commandant of the Air Ground Operations School, Nellis AFB, for his great support to this work, both in sharing his combat experiences and CAS expertise and in providing detailed comments on the draft report.

Col. Brett Williams and the staff of the Checkmate Division, Headquarters USAF, shared combat experiences and commented on the project briefing.

RAND colleagues Natalie Crawford, John Gordon, Ted Harshberger, Tom McNaugher, Rich Moore, David Ochmanek, Walt Perry, David Shlapak, Mike Spirtas, Peter Wilson, and Lauri Zeman provided helpful comments on earlier versions of this work. We thank Forrest Morgan for his contributions as a study team member at the beginning of the project.

John Stillion of RAND and Jeffrey McCausland of Dickinson University were the formal reviewers of the report. We are in their debt for their thorough, insightful, and constructive critiques.

Leslie Thornton and Natalie Ziegler provided outstanding administrative support throughout this project and jointly prepared the manuscript.

Abbreviations

AFSOC	Air Force Special Operations Command
AI	air interdiction
ALO	air liaison officer
AOC	air operations center
ASOC	air support operations center
ASOS	air support operations squadron
ATCCS	Army Tactical Command and Control System
AWACS	Airborne Warning and Control System
BAI	battlefield air interdiction
BUA	Brigade Unit of Action
CAS	close air support
CCT	combat control team
CFACC	Combined Force Air Component Commander
CFLCC	Combined Force Land Component Commander
CIA	Central Intelligence Agency
CJTF	Combined Joint Task Force
DA	direct attack
DoD	Department of Defense
DSCS	Defense Satellite Communications System
ETAC	enlisted terminal attack controller
FAC	forward air controller
FAC-A	forward air controller–airborne

FCS	future combat system
FSCL	fire support coordination line
FSCM	fire support coordination mechanism
FSE	fire support element
FSO	fire support officer
GBS	Global Broadcast System
GPS	Global Positioning System
IP	initial point
ISR	intelligence, surveillance, reconnaissance
IT	information technology
JAOC	joint air operations center
JDAM	joint direct-attack munition
JSTARS	Joint Surveillance and Target Attack Radar System
JTAC	joint terminal attack controller
KLA	Kosovo Liberation Army
LGB	laser-guided bomb
MLRS	Multiple Launch Rocket System
NATO	North Atlantic Treaty Organization
ODA	Operational Detachment Alpha
SBCT	Stryker brigade combat team
SMART-T	secure, mobile, antijam, reliable, tactical terminal
SOF	special operations forces
STS	special tactics squadrons
TAC	terminal attack controller
TACP	tactical air control party
TOC	tactical operations center
UAV	unmanned aerial vehicle
WMD	weapons of mass destruction
WP	Warsaw Pact

CHAPTER ONE
Introduction

> "No matter how bad things got for the Americans fighting for their lives on the X-Ray perimeter, we could look out into the scrub brush in every direction, into that seething inferno of exploding artillery shells, 2.75-inch rockets, napalm canisters, 250- and 500-pound bombs, and 20mm cannon fire and thank God and our lucky stars that we didn't have to walk through that to get to work."[1]
>
> *Harold Moore and Joseph Galloway*

Background

Operations against enemy ground forces and in support of friendly ground forces have figured prominently since the dawn of air power. Aviation elements partnered with ground forces in World War I, in Nicaragua in the late 1920s, and in World War II, Korea, Vietnam, Grenada, Panama, Iraq, and Afghanistan.

Airmen, particularly in the U.S. Air Force, have not always agreed with soldiers about the best way to apply the air instrument. Early on, airmen came to believe that the most effective use of air power was to strike deep against enemy sources of power—thus the emphasis on the strategic air campaign in World War II, Korea, Vietnam, Serbia, and both wars with Iraq. Strikes on leadership, communications, industry, electrical power generation, and trans-

[1] Moore and Galloway, 1992, p. 105.

portation were envisioned as war-winning by theorists from Giulio Douhet to the Air Corps Tactical School to John Warden.[2] Attacks against deployed enemy forces were generally viewed as a less effective use of air power, although interdiction was recognized as quite lethal under the right circumstances. Close support of friendly ground forces, however, was viewed by many airmen as something to be performed only under extreme conditions. Others argued that it represented a failure of air power by allowing the enemy to close with friendly forces. Airmen also viewed close air support (CAS) as wasteful, the use of a strategic asset for tactical purposes. Finally, they feared becoming flying artillery, divided up and assigned to support lower-echelon ground forces rather than exploiting air power's ability to mass and strike anywhere in a theater.

Despite these reservations, whenever U.S. ground forces have found themselves in desperate battles, airmen have come to their aid, often at great risk and with significant losses. And despite the popular conception that the Air Force as an institution does not care about CAS and interdiction, many airmen deeply believe in the mission. Maj. Gen. Pete Quesada's IX Tactical Air Force pioneered CAS and terminal control techniques in support of Gen. Omar Bradley's First Army in World War II. A later generation of airmen was bloodied flying difficult and dangerous forward air control and CAS missions in Vietnam some 20 years later.[3]

For many years, the A-10 and AC-130 flying communities, as well as Air Force special tactics squadrons (STS) and conventional tactical air control parties (TACPs) have specialized in CAS and have spent their entire careers working closely with the Army. It is true that these communities represent fairly small and, until recently, somewhat neglected Air Force subcultures. Beyond them, however, there is a growing community of fighter and bomber aircrews who, based on their recent combat experiences, have embraced the CAS mission.

---

[2] Douhet, 1983; Finney, 1992; Warden, 1989.

[3] Hughes, 1995; Harrison, 1989.

Recent operations in Afghanistan and Iraq gave unprecedented visibility to air operations against enemy ground forces. Combat controllers became better known within the defense community and, to a lesser extent, to the public for their exploits in Afghanistan, where they directed fighters and bombers to drop satellite- and laser-guided munitions on Taliban forces. Modern air power partnered with ground controllers, Special Forces, and indigenous ground forces to produce strategic effects, defeating the Taliban regime more rapidly than any had hoped. Stories of special tactics combat controllers equipped with laptop computers, satellite communications, and Global Positioning System (GPS) navigation systems, but traveling on horseback, became legendary and were prominent in speeches by the Secretary of the Air Force and the Chief of Staff.

Operations in both Afghanistan and Iraq have raised the visibility of operations against enemy ground forces (especially CAS) in the Army, the Air Force, the Department of Defense (DoD), and the joint community. The success of air power in providing day, night, adverse-weather,[4] precision support for ground forces has convinced the Army leadership that it can make its forces more deployable and agile by reducing its own artillery support (and the tons of associated ammunition, vehicles, and fuel) and relying more heavily on air power. Airmen, however, appear to have mixed feelings about this newfound Army enthusiasm for air power. On the one hand, it is a vindication of arguments airmen have been making for decades. On

---

[4] The United States has made great strides in its ability to do precision attack in adverse weather and in the use of weather forecasting to improve targeting decisions. During Operation Iraqi Freedom, Air Force weather forecasters gave planners in the Combined Air Operations Center sufficient warning to adjust targeting, weaponeering, and tactics to overcome terrible weather conditions. Although dozens of combat sorties had to be aborted or were unable to deliver weapons due to the weather, radar sensors and GPS-guided weapons enabled many aircraft to conduct successful, indeed critical, attacks in support of land forces. That said, it would be a stretch to say that "all-weather" precision support has been achieved. Severe weather limited the use of electro-optically guided systems (the most precise), hindered the battle damage assessment process, and prevented terminal air controllers (TACs) from positively identifying targets in some cases. Our thanks to Colonel Mark Wheaton and staff in the Air Force Directorate of Operations Weather division for sharing these insights.

the other hand, they remain fearful of becoming flying artillery parceled out to each company commander.

All air-land operations are inherently joint, involving the contributions of all four services. During Operation Iraqi Freedom, for example, Army battalions received CAS from Army helicopters, Air Force fighters and bombers, Navy fighters, coalition fighters, and even (much less frequently) Marine Corps fighters. However, the Army's relationship to air power is far different from that of the Marine Corps. Marine units fight as part of a Marine air-ground task force that includes attack helicopters and fixed-wing attack aircraft. The Marines train their ground and air units to fight as a combined-arms team. Their airmen see their sole mission as assuring the survival and success of Marines on the ground. Their forward air controllers (FACs) are all Marine pilots, who may also be assigned to ground units. In contrast, the Army is prohibited from developing fixed-wing attack aircraft and has no counterpart to the Marine air-ground task force. Air Force A-10 pilots, like Marine airmen, consider their sole mission to be assuring the survival and success of Army troops on the ground, but other Air Force pilots may regard air support to ground forces as just another form of strike.

Except in wartime, Army officers have little exposure to air power and little opportunity to train together with air forces. Their terminal attack controllers (TACs) are Air Force enlisted men, who are usually collocated with the units they support but are not assigned to those units. These profound organizational and cultural differences imply that the relationship between the Air Force and the Army has a peculiar character that demands special attention. However, many of the insights developed from this relationship, especially in the area of control measures, may be applicable in the broader joint arena.

Purpose and Organization of This Report

This report seeks to help the Air Force engage the Army and broader joint and allied communities in a constructive dialogue on these issues. In particular, it addresses three policy questions: (1) How should air attack and ground maneuver be integrated? (2) How should the CAS terminal control function be executed? (3) How should ground maneuver/fires and air attack be deconflicted?

Chapter Two looks at the evolving relationship between air and ground power, exploring in greater detail the doctrinal issues introduced above. Chapter Three identifies trends in counterland operations, based on an analysis of Operations Allied Force, Enduring Freedom, and Iraqi Freedom. Chapter Four considers the impact Army Transformation efforts will have on the future air-ground partnership. Chapter Five presents a quantitative analysis of the requirements associated with providing air power on-call 24 hours a day over a large battlefield. Chapter Six describes the terminal attack control function, quantifies the number of TACs that are likely to be required to support Army Transformation, and considers alternative approaches for executing the terminal attack function. Chapter Seven presents the study's conclusions.

CHAPTER TWO

The Evolving Relationship Between Air Power and Land Power

"Close-in air-ground cooperation is the difficult thing, the vital thing, the other stuff is easy."[1]

Maj. Gen. Pete Quesada, U.S. Army Air Forces, 1945

How should U.S. air power and land power be employed together on future battlefields? This chapter addresses this question by examining a range of tactical and operational alternatives and placing them in context to assess how well each meets the challenges of future military operations.

Air Power Against Armies: Counterland Operations

U.S. Air Force doctrine classifies air operations in broad functional categories: counterair, counterland, countersea, strategic attack, airlift, air refueling, and nine others.[2] Counterland operations are "conducted to attain and maintain a desired degree of superiority over surface operations by the destruction or neutralization of enemy surface forces" in order to "dominate the surface environment and [redundantly] prevent the opponent from doing the same," either in concert

[1] Scales, 1994, p. 15.

[2] The other doctrinal functions are counterspace; counterinformation; command and control; spacelift; special operations employment; intelligence, surveillance, and reconnaissance; combat search and rescue; navigation and positioning; and weather services (Air Force Doctrine Center, 2000, pp. 5–24).

with friendly ground operations or largely independent of them.[3] Counterland does not constitute all air attacks against land targets, however; combat operations in the other functional areas, especially counterair, counterspace, strategic attack, and special operations, also involve attacking terrestrial targets, including airfields, air defenses, command-and-control systems, industrial and transportation facilities, and enemy leaders.

The Spectrum of Counterland Missions

Counterland operations are directed against the ability of enemy ground forces to operate. They traditionally encompass two types of missions: air interdiction (AI) and CAS.[4] CAS is defined in U.S. Air Force and joint doctrine (and that of most major U.S. allies)[5] as "Air action . . . against hostile targets that are in close proximity to friendly forces and that require detailed integration of each air mission with the fire and movement of those forces," primarily to avoid losses to friendly fire among either ground or air forces.[6] AI, in contrast, is "air operations conducted to destroy, neutralize, or delay the enemy's military potential before it can be brought to bear effectively against friendly forces at such distance from friendly forces that detailed integration of each air mission with the fire and movement of friendly forces is not required."[7] CAS and AI are parts of a continuum, and drawing a clear line between the two is often difficult in practice. Chapter Three addresses this issue in detail.[8]

---

[3] Ibid., p. 10.

[4] For broad historical studies of these missions, see Cooling, 1990, and Mark, 1994.

[5] There is far more commonality among air forces in their definitions of these missions than in the larger doctrinal frameworks within which each situates them. See most significantly Ministry of Defence, 1999, and Royal Australian Air Force Aerospace Centre, 2002.

[6] A concise examination of the development of U.S. CAS theory and practice through the Korean conflict appears in Lewis and Almond, 1997.

[7] Joint Staff, 2002.

[8] An intermediate category called battlefield air interdiction (BAI) no longer appears in U.S. doctrine. See McCaffrey, 2002.

Strategic attack falls beyond the boundaries of counterland operations, according to doctrine, but must be considered alongside CAS and AI, since one of its major elements is attacking the resources and tools of military production and sustainment, with the goal of reducing or destroying enemy military potential on a broad scale.[9] Whether attacks against existing enemy ground forces themselves (and not just the potential to field forces) can properly be considered strategic attack is debatable.[10] An alternative approach to classifying air attack against fielded forces so remote from contact with friendly ground forces that striking them cannot reasonably be described as interdiction is to define a third counterland mission, called by its proponents direct attack (DA). The U.S. Air Force appears to have decided against this course.[11]

---

[9] See Pape, 1996. Pape calls such bombing strategies "strategic interdiction," a label that is doctrinally awkward but theoretically apt.

[10] Air Force Doctrine Document 2-1.2 now defines strategic attack as "offensive action . . . aimed at generating effects that most directly achieve our national security objectives by affecting an adversary's leadership, conflict-sustaining resources, and/or strategy." Unlike the previous version of AFDD 2-1.2 (May 20, 1998), which listed military formations that are essential for maintaining the enemy regime in power as a possible target set for strategic attack (p. 18), the new doctrine makes no mention of ground forces per se being potential strategic attack targets and presents strategic attack as an alternative to attacking fielded forces, although it does include attacking industry in order to weaken the enemy's armies, striking fielded missile forces, and in some cases interdicting military supplies under the strategic attack umbrella (pp. 11–12). U.S. joint doctrine does not currently address strategic attack at all. A detailed examination of strategic attack doctrine falls beyond the scope of the present discussion, and it is a subject that has been growing progressively more convoluted. This is strikingly demonstrated in a recent article by Air Combat Command Commander Gen. Hal Hornburg, who recasts strategic attack to include a wide variety of air operations against targets with strategic significance, including interdicting the movement of Bosnian Serb fielded forces during Operation Deliberate Force in 1995 and attacking Taliban and al Qaeda forces in Afghanistan in 2001 in concert with friendly ground operations (Hornburg, 2002). For a discussion of attacking enemy armed forces as purely strategic targets, see Mueller, 2002, pp. 117–142.

[11] When the concept of a direct-attack mission type was initially developed, its proponents used the term "battlefield air operations" to describe it. However, this label was not only awkwardly reminiscent of battlefield air interdiction (to which it was unrelated), it was also confusing in its own right, since the essence of the new mission was that it involved attacks against enemy forces still far removed from the close battle.

Differences Among the Counterland Missions

The missions across the counterland spectrum vary in a number of respects, but it is important to distinguish between those variables that differ fundamentally from one mission type to another and those that are less intrinsically related to them.

Perhaps the most obvious feature that separates CAS from AI for the casual observer is the location of the targets for each type of attack. CAS occurs near friendly forces; interdiction happens deeper behind enemy lines; and strategic attack is directed at the heart of the enemy state beyond the armies that stand poised to defend it. The types of targets that are involved also vary. CAS mainly attacks enemy combat units; AI attacks primarily softer transport, logistics, and communications assets that enable enemy forces to maneuver and fight; and strategic attack attacks industrial production capacity and other targets that enable the enemy to field, maintain, and employ its armed forces.

From these characteristics follow differences among the effects of the missions. CAS has both the most immediate and the most localized effects. AI affects a broader area of the theater, and its results take longer to be felt on the front line. Strategic attack creates far-reaching effects that are felt across most or all of the enemy's military activities but which are relatively diffuse and typically take the longest time—months or even years, when attacking the production of new weapons—to alter conditions on the front lines of a particular battlefield.[12]

The location and nature of targets for the different missions led to the evolution of different types of aircraft for each mission during the early decades of military aviation. CAS required small, agile ground-attack aircraft capable of strafing or dive-bombing point targets at relatively short ranges from their bases. AI became the domain of light and medium bombers that could carry heavier bomb loads

[12] See Olson, 1962.

longer distances. Strategic attack required heavy bombers that could range deep into enemy territory while carrying even larger payloads necessary to make such trips worthwhile.

Although this association of aircraft with missions remains fixed in the popular imagination,[13] it started to break down during World War II, particularly in the use of U.S. air power, as longer-range fighters attacked strategic targets, and heavy bombers pulverized concentrations of enemy forces on the front lines.[14] During the Vietnam War, the traditional hierarchy largely collapsed, due to aerial refueling, improving air defenses, precision-guided munitions, and other factors. Indeed, the U.S. air war against North Vietnam was conducted almost entirely by fighters and attack aircraft, while B-52 bombers concentrated on interdiction and even CAS south of the Demilitarized Zone (DMZ). Of course, aircraft are still designed and equipped with particular missions in mind: For example, the A-10 is optimized for CAS, and the F-117 is intended for strategic attack. But most attack aircraft can be employed in any counterland mission.

Traditional images of what strategic attack, AI, or CAS looks like have begun to fray around the edges in other respects. Precision weapons delivery means that strategic attacks no longer must concentrate on striking large targets, opening the door to deep attacks not only against more discrete war-supporting targets such as command-and-control nodes but also against deployed military forces. Weapon and sensor advances over the past three decades make it possible to destroy heavy enemy forces that are out of contact with friendly units even when they are not moving, though the "interdiction" label lives on. Envisioning CAS as just air power supporting ground forces is now so antiquated that it is time to consider changing the term.

What should be the primary feature distinguishing one category of counterland missions from another? Fortunately, it is the factor

---

[13] See, for example, Pape, 1996.

[14] Conversino, 1997–98. For example, in 1944, bomb-carrying P-38 Lightning fighters were used to attack the heavily defended Ploesti oil refineries, while Allied heavy bombers carpet-bombed German forces during the breakout from the Normandy beachhead.

that figures most prominently in current doctrinal definitions: the extent and nature of the coordination required between ground and air forces to avoid fratricide and to maximize the effectiveness of the various military assets. CAS differs from other counterland missions above all because of the need for detailed integration of air and ground operations to ensure with a very high degree of confidence that air-delivered ordnance will strike the correct and only the correct targets, and that aircraft will not be hit by friendly fire. AI does not require this extreme and laborious degree of coordination because friendly forces are safely out of the way, but it must be coordinated at a higher level with the scheme of maneuver on the ground, because this scheme determines which enemy ground forces and lines of communication should be attacked, when, how, and in what order. Finally, strategic air-attack targeting requires only a general degree of coordination between air and ground operations.[15]

---

[15] This does not mean that there is little relationship between strategic attack and ground operations, only that the necessary coordination resides at the strategic rather than the operational level. The overall types and timing of intended operations on the ground will fundamentally affect what strategic air operations should look like—attacking enemy steel production will be militarily irrelevant if the opposing armies will be defeated in a matter of weeks, for example, while strategic attacks against ground forces will have greater impact if the enemy anticipates facing an invasion in the near future than they will if his main challenge is to survive a blockade.

It also does not mean that the absence of friendly ground forces makes it unimportant to deliver weapons precisely and discriminately. Most obviously, the proximity of civilians to the targets being attacked may necessitate as much care and restraint in an attack as would be required if friendly military forces were nearby. However, in such cases, this tends to be a problem for the air force in question to solve on its own, since there is likely to be no means of interactively coordinating its attacks with the activities of the civilians in the same sense that coordination occurs in CAS. The attacks may still be conducted with close attention to the behavior of the noncombatants—avoiding attacks against bridges during periods of peak traffic, for example, or broadcasting warnings for civilians to stay clear of target areas—but such decisions will be made independently by air commanders. Even in strategic attack, close air-ground coordination is required at the tactical level on those occasions when friendly special operations forces (SOF) are present to designate or identify targets.

Operational Conceptions of Air Power and Land Power

While this study primarily concerns the tactical issues associated with CAS, it also explores the evolving relationship between air power and land power at the operational level of war, where it offers a set of more abstract, even philosophical, approaches to thinking about the relationship between land power and air power.

This section presents such a typology in order to place subsequent discussion of these issues into a useful theoretical context. It divides many possible points of view on the air-land relationship into a spectrum of five categories (summarized in Table 2.1), describing them in order from the most ground-centric to the most air-centric. Inevitably, each category encompasses a number of disparate theories and doctrines, and the boundaries between them are artificial. Finally, these are general perspectives on the predominant nature of the rela-

Table 2.1
Perspectives on the Air-Land Relationship

Relationship of Air to Land Power	Air Augments Land	Air Complements Land	Air Partners with Land	Air Dominates Land	Air Trumps Land
Supported force	Land	Land	—	Air	Air
Typical air missions	CAS and AI	AI	AI and direct attack	Direct attack	Strategic attack
Key air-power contribution[a]	Flying artillery	Shaping the battlefield	Hammer and anvil	Death from above	Going downtown
Likely future appropriateness	Occasional	Sometimes, especially against unconventional forces	Frequent	Sometimes, especially against conventional forces	Infrequent

[a] These labels are deliberately colloquial and provocative. "Flying artillery" implies a concept that would treat aircraft like indirect-fire weapons. "Going downtown" implies attacking targets of strategic importance other than fielded forces. In all cases, air superiority is prerequisite.

tionship between air power and ground power, and no single point of view will apply to every set of circumstances.

Air Power Augments Land Power

In the first perspective, counterland air power provides additional fires to supplement those of friendly land forces in the close battle. It contributes to victory in the decisive close battle through CAS and AI. From this perspective, the effects of aerial fires are not fundamentally different from those of land-based fires, particularly artillery fire, and the relationship between the two is one of fairly straightforward substitution. Aerial firepower offers particular advantages relative to artillery and rockets, but terrestrial firepower could compensate for its absence.

Substitution may be intermittent, with ground forces calling on air power to fill temporary firepower shortfalls during intense combat or other emergencies, or long-term. The former is well illustrated by the use of air power to help destroy frontline Iraqi units in southern Kuwait and Iraq prior to the start of the coalition ground offensive in Operation Desert Storm, when thousands of attack sorties were added to the weight of a massive artillery bombardment to clear the way through the Iraqi defenses. Perhaps the most obvious example of long-term substitution is the U.S. Marine Corps's use of attack aviation. Although Marine aviators also perform AI, CAS is their *raison d'être*, and the Marine Corps explicitly invests in air power to compensate for having substantially less artillery and armor than heavy forces in the Army have.[16] Aircraft serve the Marines well as a partial substitute for artillery, not least because they can be carrier-based during amphibious assault. Similarly, when the U.S. Army first fielded armed helicopters to support air-mobile operations in Vietnam, they were officially designated as "aerial rocket artillery."[17]

---

[16] In particular, the Marines have no equivalent of Army corps-level artillery, relying on air power to augment their artillery assets at division level and below.

[17] Scales, 1994, pp. 92–93.

This image of air power as "flying artillery" is anathema to airmen, who argue that air power makes a greater contribution by reaching across the theater and deep into the enemy rear. It threatens to squander this potential by employing air power in small elements shackled to local ground operations. The struggle against this perspective has been the principal theme in the development of U.S. Air Force doctrine for many decades.

Yet airmen should not unthinkingly reject the "flying artillery" perspective simply because of a dogmatic reflex. There are circumstances in which this is the optimal way to employ air power, and its effectiveness in this role has increased as sensors and precision munitions have improved. Enabling air power to perform up to its potential in this capacity without constraining its ability to make other contributions to joint warfare is a challenge that subsequent chapters of this report will address.

Air Power Complements Land Power

The second perspective also starts from the premise that friendly land forces ultimately play the central role in defeating the enemy, but it sees aerial firepower affecting the enemy in ways that land forces cannot duplicate, although land forces will ultimately deliver the *coup de grace*. Because air power can strike enemy forces at depths far beyond the range of tube artillery (typically 30 to 40 kilometers), this perspective emphasizes affecting the close battle through AI, weakening or immobilizing enemy forces before they can maneuver against friendly units so that they can be defeated in detail on the ground. AI does not eliminate the need for CAS, but it generally does far more to favorably shift the balance on the battlefield.

Such an approach to the employment of air power was prominently championed in John Slessor's seminal *Air Power and Armies* in the 1930s, as an alternative to both the "flying artillery" mindset and single-minded emphasis on strategic attack.[18] The German use of air power in *Blitzkrieg* operations at the start of World War II followed

---

[18] Slessor, 1936; Meilinger, 1997, pp. 41–78.

such a model,[19] and it has arguably been the most prominent approach to counterland air power in most major conflicts since that time, including the ground combat phase of the 1991 Persian Gulf War. Air power as a complement to ground power was perhaps most visibly embodied in the AirLand Battle doctrine of the 1980s, when air power's primary contribution to defeating a Soviet-led offensive in central Europe was to have been interdiction of second- and third-echelon Warsaw Pact (WP) forces. Through attrition and by inhibiting the movement of these follow-on forces, air power was expected to prevent the WP from massing the forces necessary to break through the defenses of the North Atlantic Treaty Organization (NATO).

The evolution of U.S. Army attack-helicopter doctrine helps illustrate the difference between this perspective and that of "flying artillery." While the U.S. Marines have consistently emphasized the use of attack helicopters in small elements as part of the combined-arms team, most Army attack aviation emphasizes larger-scale deep attack missions in preference to close-combat attack, with battalions of attack helicopters penetrating deep into hostile territory to strike *en masse* against enemy force concentrations. Advocates argue that a few large-scale deep attacks will have more impact on the enemy than a larger number of smaller attacks in close combat will. The Air Force's traditional preference for strategic attack and AI over CAS is based on the same rationale. It remains to be seen whether recent experience in Afghanistan and Iraq will cause the Army to change this emphasis.

Air Power Partners with Land Power

While the first two perspectives are frequently represented in works of military theory and in the actual employment of air power over the past 90 years, the third and fourth perspectives are relatively new, growing out of the increasing ability of air forces to destroy enemy ground forces under a wide range of conditions, a trend that will be

---

[19] Van Creveld, Brower, and Canby, 1994, Chaps. 2 and 3. Indeed, the Luftwaffe's training, equipment, and doctrine were optimized for AI to such a point that it was ill-prepared to carry out missions such as strategic attack, countersea operations, and airlift.

discussed in greater detail later in this chapter. The third perspective, which calls for a synergistic partnership of air power and land power, is the least familiar and the most important for this study.

This perspective involves more than merely distributing responsibilities in an evenhanded way; it is not just a compromise. Instead, it takes the view that air power and land power offer different but complementary advantages and limitations to its logical conclusion. It suggests that with the maturation of air power, it is now possible to envision air and ground power operating in a mutually reinforcing hammer-and-anvil relationship, where either might do the greater amount of damage, depending on the situation. In this "double attack" relationship, air power will wreak havoc against large concentrations of enemy ground forces, confronting them with a dilemma: If they concentrate, air forces will destroy them. If they disperse, land forces will overwhelm them.

Thus the air-land relationship from this perspective is one of relative equality at the aggregate level, while at the tactical level, the campaign is likely to be marked by frequent and often unpredictable alternation between fights in which air power supports a ground-centric scheme of maneuver and those in which the reverse is true. Because this approach to warfare is neither air- nor ground-centric, embracing it is doctrinally challenging for organizations that are steeped in a tradition of command relationships that designate component commanders as either supported or supporting.

Historical examples of air-land partnerships are not abundant. Perhaps the best, yet imperfect, example is Operation Iraqi Freedom in spring 2003. As U.S. Army and Marine forces drove toward Baghdad, there was a pattern of relatively rapid handoffs of the leading role in attacking the Iraqi army, with air power mauling concentrations such as the Medina Division in place and ground forces taking the lead, frequently with support from air power, in destroying those Iraqi forces that moved or stood against them in smaller numbers. Similar effects might have occurred had Operation Allied Force not impelled Serbia to capitulate and had NATO intervened with land forces in Kosovo during late 1999. NATO air attacks against Serb forces in Kosovo achieved little, due in large part to the lack of sufficiently effective land forces to serve as an anvil for the hammer of air power.

Air Power Dominates Land Power

The fourth perspective takes an air-centric approach to counterland operations. It suggests that because air power can attack ground forces with great lethality while minimizing the risk of friendly casualties, air forces should be the principal instrument of destruction, while land forces facilitate this role. The principal mission in such operations would be what internal U.S. Air Force discussions have called "battlefield air operations" or, more recently, "direct attack," i.e., air attacks against enemy ground forces in which friendly land forces either support the air attacks or are absent altogether. These attacks differ from AI in that the planned air operations are the principal determinant of where and how (or even whether) land forces will be used, instead of the other way around.[20]

Such an air-dominant approach can take several different forms. One is for air power simply to destroy enemy ground forces on its own, as is envisioned in the U.S. Air Force's concept of the "halt phase": countering an enemy invasion of a weakly defended ally by an intense air campaign that would stop the invasion before the enemy could achieve his objectives. If necessary, friendly land forces would arrive and subsequently expel the invading forces from any territory they had conquered. A variation on this theme would employ friendly ground forces in small, light units, such as special operations forces (SOF), to spot and designate targets for air attack, relying on air power to provide the heavy firepower needed to destroy the enemy forces. Operation Enduring Freedom and Operation Iraqi Freedom (in northern and western Iraq) offer successful examples of this approach.

Another way in which ground forces might facilitate an air-centric counterland effort would be to deploy a ground force to threaten the possibility of a land offensive in order to force enemy

---

[20] Such air attacks might also be characterized as strategic attack if the definition of strategic attack does not exclude enemy land forces as targets.

units to adopt defensive positions that would increase their vulnerability to air attack.[21] This happened during the six-week air campaign of Operation Desert Storm. The Iraqi army was damaged to such an extent that Saddam Hussein might have agreed to withdraw from Kuwait if doing so would have allowed him to preserve his remaining forces, although he had previously rejected such an offer.[22]

Air Power Trumps Land Power

At the air-centric extreme of the spectrum is the view that air power can achieve such decisive effects against an enemy's war-making potential that armies become irrelevant. This was the perspective of strategic attack purists of the 1920s and 1930s, the most prominent among them being Giulio Douhet. He held that air forces could attack the enemy's center of gravity so effectively that the enemy state would quickly collapse, and thus that air forces alone could achieve victory.[23]

Douhet-like prescriptions for strategic bombing, which emphasize inflicting destruction on civilian populations, have largely been abandoned by Western nations since World War II. However, belief in the omnipotence of strategic attack lives on in air-power theories such as that of Col. John Warden. According to these theories, intensive strikes on enemy leadership, infrastructure, command-and-control systems, and other nonmilitary targets can produce victory, leaving friendly land forces little to do other than protect airbases and occupy territory after the enemy has capitulated.

Such strategies have had little success in the past,[24] and in 1990 Warden's argument that the potential of strategic air attack made Iraq's army irrelevant met a distinctly cool reception from his superiors in U.S. Central Command, although they did conduct strategic

---

[21] Land forces might assist an air-centric strategy by keeping enough pressure on enemy forces to make them consume resources that could no longer be replaced because of air attacks. But in this case, friendly land forces would likely play an increasingly important role.

[22] Pape, 1996, Chap. 7.

[23] Douhet's major works are collected in Douhet, 1983. See also Meilinger, 1997, pp. 1–40.

[24] Pape, 1996; however, see also Mueller, 1998, regarding other effects of strategic air attack.

air attacks against Iraq substantially along the lines he had advocated. The debate recurred during Operation Allied Force in 1999, when NATO's air-component commander, Lt. Gen. Michael Short, argued that Alliance air power should concentrate on strategic targets in the Serbian heartland rather than the fielded forces in Kosovo as demanded by Gen. Wesley K. Clark, the theater commander. In the end, both avenues were pursued, and whether or not Serbia's eventual capitulation was heavily influenced by an expectation that NATO would eventually invade or other political factors, the air campaign against Serbian strategic targets was certainly more powerful in determining the conflict's outcome than was the counterland effort, which proved to be largely ineffective at destroying dispersed and camouflaged ground forces in Kosovo.[25]

The Air-Land Partnership in Perspective

Of these five broad perspectives, the third, implying integration of air and ground power at the operational level, is rhetorically very appealing. Jointness, at least as an ideal, is difficult to dislike. But two questions need to be addressed before deciding to embrace partnership. The first question is whether such integration is feasible. The obstacles are substantial. The services would have to step away from the tradition of defining command relationships in terms of supported versus supporting commanders. Moreover, true integration requires operational commanders who are well versed in air, land, joint, and combined operations, a breadth of training, education, and experience that few officers possess.[26] The second question is whether moving in this direction is worth the effort; this question is addressed in the next section.

---

[25] On the factors that appear to have influenced Milosevic's decision to capitulate, see Byman and Waxman, 2000; Hosmer, 2001a; Lambeth, 2001.

[26] See Belote, 2000.

Envisioning Air Power and Land Power on Future Battlefields

Given the range of possible relationships between air power and land power and considering historical trends, what should the relationship be on future U.S. battlefields? The remainder of this chapter attempts to answer this question in broad terms. Chapter Three will look in greater detail at experience during recent counterland operations in Serbia, Afghanistan, and Iraq.

Counterland Operations Are Critical to U.S. Strategy

Impassioned advocates of air power typically do not speak or write a great deal about counterland air attack. With some exceptions, air superiority and strategic attack have been the dominant themes in seminal books of air-power theory and both scholarly and popular accounts of the use of air power during the past century.[27] Yet the destruction of enemy armies has usually been a necessary condition for military victory even in the age of air power and has often been sufficient to achieve it.[28]

Counterland attack is not the only way air power can contribute to the defeat of enemy ground forces—achieving air superiority is usually of critical importance to events on the battlefield below and is generally a prerequisite for effective counterland air operations, while strategic attack may do much to weaken the enemy's ability to fight. However, less-glamorous air attack against fielded forces is frequently central to military success, especially for the United States. Air interdiction and CAS made the liberation of Western Europe possible

---

[27] Air superiority is an absolute precondition for successful joint operations. However, airmen's enthusiasm for air superiority and strategic attack also reflects the fact that air power operates most independently of land forces in these roles. Emphasizing strategic attack once also served to maximize budgets and force size, since air forces had to be large and well equipped to mount powerful raids deep into enemy territory. Today the situation has changed to some extent: When a few stealthy aircraft armed with precision munitions can achieve the strategic attack effects of hundreds of old-fashioned heavy bombers, and when few potential adversaries present daunting air-to-air threats, counterland operations have become the most powerful justification for a large air force.

[28] Pape, 1996; Mueller, 1998.

during World War II. In Vietnam, air power was key to halting the 1972 Easter Offensive, the great U.S. success of the war. Counterland air power crippled the Iraqi army in 1991, rapidly turned the tide in Afghanistan in 2001, and mauled the Iraqi army again in 2003, each time leading to decisive results. The 1999 defeat of Serbia was an exceptional departure from this pattern.

In some contingencies, enemy fielded forces may be the only politically acceptable ground targets against which air attack can be usefully applied at all or the only targets meriting large-scale air attacks.[29] Such contingencies are likely to figure increasingly prominently in U.S. military operations in the future, as the United States seeks to deal with active and potential terrorist threats and as the number of opponents willing to use conventional military means against the United States dwindles.

Attacking fielded military forces also tends to appeal to statesmen and strategists because of its political and legal acceptability. The laws of warfare prohibit attacks against purely civilian targets, and even attacks on dual-use targets may entail significant political costs, a problem that becomes more acute in coalition warfare, when multiple states must concur. In contrast, attacking military forces is unambiguously legal and usually appears to be both moral and legitimate, making leaders far more willing to authorize it and to delegate managing its execution to military commanders.[30]

Air Power Is Increasingly Effective Against Land Forces

The ability of air power to destroy land forces has increased from generation to generation as aircraft, sensors, and ordnance have improved. Air power was already a mighty force against armies on the battlefields of World War II, to the point where possessing effective air superiority was often a necessary condition for tactical and operational success. Yet its counterland capabilities were still limited in many respects. Ground forces were largely immune from air attack at

---

[29] See Corum and Johnson, 2003.

[30] Even in attacks against military forces, moral considerations may arise, however, particularly when the personnel being attacked are unwilling conscripts of a dictatorial regime.

night or in poor flying weather, for example. Armored vehicles could be engaged from the air but were difficult to destroy. And to strike ground forces with any degree of accuracy required aircraft to close to ranges at which antiaircraft defenses often inflicted loss rates that would be considered horrific by contemporary Western standards. Even under favorable conditions, effects against ground targets were typically on the order of multiple, and often many, attack sorties per kill.

Today the conditions look quite different for modern air forces, though air power still has limitations as a counterland instrument. Night attack capabilities have become widespread with the development and proliferation of improved sensors—indeed, targets are sometimes more vulnerable at night than they are during daylight, and adverse-weather strike capabilities have been greatly enhanced by the advent of satellite-guided munitions.[31] Precision-guided weapons have made attacking point targets such as armored vehicles much more effective, and they enable aircraft to mount effective attacks from altitudes and ranges that reduce overall losses from most antiaircraft defenses to levels that would have been inconceivable 60—or even 20—years ago.[32] Precision-guided munitions also allow highly discriminating attacks at a strategic level. Finally, a far less conspicuous revolution in surveillance and battle management is further increasing U.S. ability to direct aircraft and weapons against targets not only effectively, but also efficiently and quickly.

Of course, land forces have become more destructive as well. The close battlefield has become a more dangerous place due to im-

---

[31] However, particularly inclement weather can still prevent air operations, a consideration that should be borne in mind when balancing reliance on air- and ground-delivered firepower.

[32] The very low loss rates among U.S. aircraft in recent wars occurred when operating against far weaker adversaries whose air defenses fell very far short of being state-of-the art. Fortunately for the United States, highly advanced surface-to-air missile systems have proliferated substantially more slowly in the years since Operation Desert Storm than many observers expected. Now that fixed-wing aircraft are highly effective against targets such as armor formations, the rationale for employing helicopters for the deep attack mission has become highly questionable, particularly in the wake of events in Kosovo, Afghanistan, and Iraq that will be discussed in the following chapter.

proved surveillance, more-precise aiming, and ever-more-deadly weapons. Indirect fire has become more deadly because of counter-battery radar and the development of area-effect submunitions.[33]

Air power remains far from omnipotent, especially when dealing with targets that are difficult to detect or identify and in urban or other complex terrain.[34] Yet its counterland effectiveness has increased to the point where it has often become possible to speak in terms of kills per sortie instead of sorties per kill, even against armored vehicles, at least when attacking conventional ground forces. This is not simply a matter of inexorable technological progress, akin to observing that automobiles have become progressively more sophisticated and reliable over the years. Instead, there are two relative considerations that matter here. First, the ability of air power to attack ground forces has increased faster than has the ability of the latter to survive in the face of air attack. Second, the ability of air power to destroy ground forces has increased more rapidly than has the ability of ground forces to kill other ground forces.[35]

These developments have caused a fundamental shift in the roles that air power can play against enemy ground forces. Military formations deep in enemy territory can be destroyed even when dug into defensive positions. Highly effective AI can reduce the need for CAS by preventing major enemy attacks and counterattacks, while the same developments make CAS a far more effective force on the battlefield than it once was. New doctrine and tactics need to take these changes into account.

---

[33] Scales, 1994, p. 113.

[34] See Vick et al., 2000, 2001.

[35] This should not come as a surprise, considering the baseline being used for the comparison. For example, in the 1940s, modern artillery had been developing for the better part of a century, to the point where it was the dominant killer on the battlefield, while air power was still relatively young. It would have been quite remarkable if air power had not made proportionally greater progress in the years that followed. Thus, this assessment should not be taken to imply, for instance, that the designers of ground-force weapons and doctrine have been less clever or innovative than their air-power counterparts.

Land Forces Provide Unique and Essential Capabilities

The increasing capabilities of air power relative to land power do not, however, imply that land forces are becoming obsolescent. Land forces can perform some tasks that air forces simply cannot, and they can perform some others far more effectively or efficiently than air power can.

The tasks that can be performed only by land forces are particularly those that involve human contact with the local population. There are a vast number of such tasks, ranging from search-and-destroy missions against guerrillas to policing occupied territory, from collecting human intelligence to performing all manner of constructive activities, either to win the hearts and minds of the beneficiaries or simply for humanitarian purposes. That air power cannot do these things (though it often supports those who do) is so obvious as perhaps to appear not worth mentioning, yet such tasks are often central to military operations, so that there is no question of air power operating alone. Activities requiring human contact tend to be most critical in counterinsurgency, stabilization, peacekeeping, "nation building," and related military operations, missions that have become increasingly important in U.S. strategy since 1989 and that are likely to predominate for the foreseeable future.

Aside from tasks that require human contact, land forces have comparative advantages over air power, especially fixed-wing air power, in a number of other areas. Most significant among these is delivering firepower quickly against suddenly emerging targets, as in the case of counterbattery fire, whenever friendly artillery has sufficient range. When fast reaction times are required, normally on the order of two minutes or less, artillery will generally be more satisfactory than air power unless an aircraft can be placed on station before a target emerges, for example, when an AC-130 flies a protective orbit or an attack helicopter provides overwatch for ground units. During the U.S. counterinsurgency campaign in South Vietnam, the mission of delivering quick-response fires to support patrols encountering guerrilla ambushes was progressively taken over by a system of artillery firebases throughout the country instead of maintaining aircraft on patrol to provide CAS on short notice.[36] Having land forces in close proximity to the enemy is often essential to locate and designate targets as well. Thus, land forces make a uniquely important contribution to effective use of air attack. In summary, land forces are usually indispensable and offer certain capabilities that air power cannot expect to replace.

Land Forces Are Increasingly Reliant on Aerial Firepower

To increase the strategic and tactical mobility of its forces and to reduce their logistics demands, the U.S. Army has begun changing its force structure to place greater reliance on medium-weight forces that are more vulnerable to enemy fires than traditional heavy forces are, and it is reducing the weight of organic artillery fires more generally. Medium-weight forces compensate for the reduced mass of fires and weight of armor by investing in systems that provide enhanced situational awareness, by deploying weapons with greater precision, and by taking advantage of the firepower that air power can provide. They are also intended to maneuver more quickly and to operate dispersed over far greater areas than in traditional ground-combat doctrine, with a less well-defined front line separating friendly and enemy forces, a concept that manifested itself in the 2003 invasion of Iraq.[37]

At the time of this writing, it remains unclear how far Army Transformation will extend and what form it will ultimately take, but

---

[36] Scales, 1994, Chap. 3. Of course, this depends on the artillery being in position to provide such fires, which can be problematic in high-speed maneuver warfare or highly dispersed operations.

[37] A degree of caution is in order, however, regarding how much change to anticipate in the nature of terrestrial battlefields and military operations. For example, it seems highly unlikely that the United States would move beyond dispersed military operations into an approach at the operational level that would be so decentralized as to properly be described as swarming. The basic rationale that favored the development of hierarchical military organizations has not changed. Above the very tactical level, swarming remains a technique of the weaker force. Indeed, even the extent to which the units of the U.S. Army's Units of Action would actually be dispersed in practice may be far less extreme than is projected in current official visions. In short, although the trends discussed in this paragraph can be expected to have a significant impact on future battlefields, it is highly unlikely that they will change matters enough to make 20th-century military experience irrelevant.

the process is well under way, as demonstrated by the establishment of the first Stryker brigades. The operational effectiveness of the Army's evolving force structure will depend heavily on how well the new forces work together with the Air Force. Moreover, analogous transformation is occurring in the armies of a number of U.S. allies, particularly with respect to emphasis on motorized forces. In combined operations, these forces, too, will depend more than ever on air power, often U.S. air power, for firepower and protection, making the issues addressed in this study important for NATO and allied doctrine and training, as well as for that of U.S. forces.

Enemy Reactions Reinforce the Need for Air-Land Integration

Potential enemies are well aware of the trends we have discussed. In particular, the enormous destructive potential of U.S. air power has prompted them to adopt countermeasures intended to offset this advantage. At the tactical and operational levels, enemies will try to prevent U.S. forces from bringing their firepower to bear, using one or both of two basic approaches.[38] The first is to avoid attack through concealment, based on some combination of camouflage; deploying in small, light, or dispersed units; and operating in complex terrain, especially cities. The second is to deploy and fight so close to civilians and friendly forces that U.S. firepower, especially air-delivered ordnance, rockets, and other artillery, cannot be used freely or even safely.

One effect of such countermeasures will be to increase the need for U.S. or other friendly land forces to flush out enemy forces.[39] If enemy forces remain unobserved until they open fire, presumably expecting to do some damage, there will be few opportunities to conduct air attacks that do not require very close coordination with land

[38] Other countermeasures, such as constructing deeper and harder bunkers to protect key strategic assets, are likely. An enemy may also adopt antiaccess measures to disrupt the ability of the United States to launch attacks in the first place.

[39] Another role of U.S. ground forces is the suppression of enemy air defenses that cannot adequately be dealt with from the air—for example, using long-range rocket systems to attack surface-to-air missile sites.

forces. Moreover, if the strategic objective of such tactics is to inflict unacceptable U.S. casualties, providing protective firepower becomes an extremely high priority. In short, adversaries have very strong incentives to fight U.S. forces in ways that maximize the need for integration of air and land forces.

Why Forge a New Air-Land Partnership?

Taken together, these trends strongly suggest that of the five perspectives defined earlier in this chapter, the air-land partnership concept is best suited to evolving strategic conditions.

The enduring—and, on the whole, growing—importance of counterland operations in U.S. military strategy does not render the strategic attack mission unimportant, but it does prevent the realistic planner from relying heavily on any approach to warfare that treats attacking enemy fielded forces as irrelevant. There may well be situations in which victory can best be achieved using such strategies, and airmen should be prepared for such contingencies, but they are likely to remain the exception rather than the rule.

At the other end of the spectrum, the increasing lethality of air power against armies does not imply that the Air Force should never be used to augment land forces' firepower, but it does mean that under most circumstances, the "flying artillery" approach to air-ground integration will not maximize air power's counterland capabilities.[40]

None of the three more-balanced perspectives is precluded to the same degree by likely future conditions. However, the second and fourth perspectives are likely to be less useful than the perspective of partnership. Opportunities for air power to operate primarily in the direct-attack mode, occasionally supported by small deployments of land forces, will be limited by enemy countermeasures such as dis-

---

[40] This does not imply that the U.S. Marines' approach to the use of air power has become invalid. Marine doctrine is not intended to be universally applicable across the joint force. It does not imply that all air power should be employed in this fashion, only that Marine air power should be.

persing forces and intermingling with civilian populations. The ability of air power to dominate land power will also be affected by the nature of the conflicts being fought, with counterinsurgency, peace operations, and other lower-intensity warfare calling for land forces to play such a major role that subordinating them to air power will rarely be appropriate. On the other hand, the Army's transformation to lighter, more-agile forces operating in fluid fashion will make air-delivered fires so important both offensively and defensively that using air forces merely to support a ground scheme of maneuver would diminish the effectiveness of the air-land team.

The third perspective, air-land partnership, will certainly not be applicable to all future joint operations, but it stands out for two reasons. First, it is appropriate to a large proportion of future adversaries and could be easily adjusted toward greater roles for either land or air power. Second, it has received relatively little concentrated attention from military theorists or practitioners, because of difficulties inherent in realizing it and because it does not assert the preeminence, and therefore serve the parochial interests, of any military service.

Partnership would not, of course, imply having co-equal commanders of the same operation, thus violating unity of command. It would imply an allocation of authority that maximized the contributions of each partner toward a common endeavor. Within the range of his organic weapons (normally 30 to 40 kilometers), the land-force commander rightly expects to control air attacks. Indeed, he must have such control in order to integrate direct fires, artillery, rockets, attack helicopters, and fixed-wing aviation. Beyond that range, an air-force commander should control air attacks, but with a view to assuring successful maneuver of land forces. In either case, the ultimate goal is not the successful employment of air power or the successful maneuver of ground forces. These are both subsidiary goals that contribute to the larger objective of destroying enemy forces. Thus, neither of these commanders need be considered supported or supporting according to doctrine, since both work for the same joint-force commander.

It would be hyperbole to suggest that the U.S. armed forces *must* embrace and develop such an air-ground partnership. The United

States is so far superior to most potential enemies that it can usually employ its military capabilities in less-than-optimal ways without jeopardizing its chances of military victory. Instead, developing an air-land partnership should be seen as an important opportunity to enhance U.S. military capabilities. It is also a challenging one, but as with all lucrative opportunities, failing to accept it would entail substantial costs, in this case potentially measured in both reduced military effectiveness and friendly casualties.

The subsequent chapters of this report address various aspects of this challenge and propose ways in which stronger integration of air power and ground power can be forged. They consider the subject at both the operational and the tactical level, and they deal with matters of doctrine, organization, and equipment. The greatest attention is devoted to the problems of applying aerial firepower close to friendly troops in contact with the enemy, for Gen. Quesada's wartime observation with which this chapter began still applies and is perhaps more true today than ever.

CHAPTER THREE

Trends in Counterland Operations

Introduction

This chapter draws insights for counterland from conventional operations during Operation Allied Force (Kosovo), Operation Enduring Freedom (Afghanistan) through March 2002, and Operation Iraqi Freedom (Iraq) through August 2004. It concludes with proposed changes to doctrine that would facilitate counterland operations in the future.

Since Operation Allied Force, NATO forces have remained in Kosovo, where they enforce an uneasy peace. The goal of a democratic, harmonious, multiethnic Kosovo is still beyond reach and may never be attained. Operation Enduring Freedom has already lasted almost three years and seems destined to continue even longer. U.S. forces are still searching for al Qaeda and remnants of the Taliban, but these adversaries are extremely hard to find. Regional leaders or "warlords," not the interim government, control most of Afghanistan. Operation Iraqi Freedom removed the Ba'athist regime of Saddam Hussein but failed to prevent a breakdown in civil order and widespread looting. Pacifying Iraq, whether as an occupation force or in support of the interim government, has proven far more difficult and risky than deposing Saddam Hussein was. Except for the Kurds, who enjoy substantial autonomy, most Iraqis have long since tired of U.S. forces, but active resistance is concentrated in Baghdad and the Sunni Triangle, and in a few Shi'ite cities in the south. The United States cannot resolve this conflict militarily, and success will depend upon the emergence of an Iraqi government strong enough to assure order.

In all three cases, conventional military operations were merely the precondition for ultimate success, which will depend on many other factors, both civilian and military. In Afghanistan and Iraq, military operations have increasingly assumed the character of counterinsurgency mingled with counterterrorism. Air forces are making important contributions to counterinsurgency in Afghanistan and Iraq, but more in the areas of transportation, surveillance, and reconnaissance than in strike, which is required only episodically. When air forces do release ordnance, they usually engage critical fixed facilities (safe houses, weapons caches, etc.), high-value targets (opposition leaders), and opposition forces in contact, often at very short distances from friendly forces. These ongoing operations will yield important insights for the Air Force, but they lie outside the scope of this research.

Land Forces Are the Critical Target Set

In practice, World War II often approached total war, implying that an enemy's entire war-making potential was a legitimate target. In Europe, U.S. and U.K. air forces destroyed the cores of cities, either incidentally to destruction of other targets or deliberately in an attempt to force capitulation. But since World War II, the United States has limited its wars and constrained use of its air power, even in the conflicts in Korea and Vietnam. Since the end of the Cold War, this tendency has become more pronounced. The current "war on terrorism," for example, demands highly precise targeting of the terrorists to avoid incidental death and destruction that would discredit U.S. leadership and alienate its allies. To meet this demand, newer weapons tend to be more precise and to have smaller radii of effect. For the foreseeable future, the United States is likely to minimize civilian loss and concentrate on military targets, especially the enemy's ground forces.

The U.S. constrains air attacks for reasons of national character and policy as well as respect for international law. Out of common humanity, U.S. leaders avoid attacking civilian targets to the extent compatible with the safety of U.S. forces. Even were U.S. leaders inclined to be more ruthless, their political goals would still demand restraint. The United States could not expect to stop barbarism in Kosovo by conducting barbaric attacks on the Serb people. Indeed, too much collateral damage could have broken the consensus within NATO, giving Milosevic his best chance at victory. During recent operations in Afghanistan and Iraq, the United States sought to replace dangerous regimes with governments that would be more peaceful and democratic. Prior to invading Iraq, the United States was already planning reconstruction and therefore sought to minimize damage to the Iraqi infrastructure. The United States intended to remove the Ba'athist regime, not to punish Iraqi citizens for its misdeeds.

While civilian targets are increasingly withheld, no such compunction applies to enemy forces. Serb military and police forces in Kosovo, Taliban militia in Afghanistan, and Iraqi armed forces of all descriptions were legitimate targets. Moreover, their destruction was key to success. Destroying Serb ground forces in Kosovo would have stopped the brutal oppression of Kosovar Albanians, a key goal for the NATO allies. Taliban rule could not survive the loss of its ground forces. Nor could Saddam Hussein's neo-fascist regime survive the loss of its various military and internal security forces. Enemy ground forces are a target set that is legitimate and usually critical to success. The challenge is to find these forces and to attack them before they can hide, without inflicting unacceptable levels of collateral damage. Table 3.1 lists counterland effects that were crucial to success in Kosovo, Afghanistan, and Iraq.

Joint Action Is Improving Counterland

During World War II, air power could be devastatingly effective against enemy ground forces, but only under the right circumstances: daylight, clear weather, and enemy forces moving or otherwise revealing their positions. And to be effective, pilots had to fly their air-

Table 3.1
Counterland Effects in Three Conflicts

Conflict	Strategic Goal	Operational Objectives	Desired Effects of Air Attacks
Kosovo	Deter offensive against the people and damage the Serb military's capacity to harm the people of Kosovo[a]	Neutralize Serb forces able to threaten the civilian inhabitants of Kosovo	Interdict the movement and supply of Serb forces in Kosovo Destroy Serb regular Army and police forces in Kosovo
Afghanistan	Disrupt the terrorist base of operations; attack the military capability of the Taliban[b]	Disrupt terrorist organizations and degrade Taliban forces	Destroy small groups of al Qaeda Protect U.S. SOF and indigenous opposition leaders from attack Destroy the Taliban's fielded forces
Iraq	Disarm Iraq, free its people, and defend the world from grave danger[c]	Defeat the forces supporting Saddam Hussein's regime	Destroy regular forces in operational depth Prevent regular forces from maneuvering Destroy paramilitary forces and militia in contact with friendly forces

[a] President William Jefferson Clinton, "Address to the Nation," The White House, Washington, DC, March 24, 1999.
[b] President George W. Bush, "Presidential Address to the Nation," The Treaty Room, Washington, DC, October 7, 2001.
[c] President George W. Bush, "President Bush Addresses the Nation," The White House, Washington, DC, March 18, 2003. For uniformity, the source in each case is the initial Presidential Address to the Nation. However, the administration's goal in Afghanistan expanded from attacking Taliban capability to overthrowing the Taliban regime.

craft at the enemy formations, making themselves targets for enemy gunners. Today, U.S. air forces can attack at night and during bad weather, even during the severe sandstorms encountered in Iraq. They can usually deliver ordnance at acceptable risk to themselves, above the range of small arms and light antiaircraft artillery. However, detecting and identifying even heavy enemy forces remains a difficult problem if they are free to hide, as they were in Kosovo. The greatest increase in effectiveness is achieved in combination with friendly ground forces that compel enemy forces to reveal themselves and help target them. Friendly ground forces may range from lightly armed indigenous forces accompanied by U.S. special operators, as in Afghanistan, to heavily armed U.S. conventional forces, as in Iraq.

In the past, delivery of ordnance near friendly forces was constrained by a lack of situational awareness and inaccurate weapons delivery. TACs might have only a very general idea of enemy locations and no way of designating those locations more precisely than with smoke or reference to landmarks. Worse yet, locations of friendly forces could be uncertain, especially in fluid situations. On top of these uncertainties, weapons delivery was subject to errors much greater than the radii of effects. To prevent fratricide, distance safe[1] had to be fairly generous, allowing the enemy to gain respite from air attack by hugging U.S. forces. But recent advances in technology have shrunk this space to a few hundred meters, less than the range of small arms.

TACs have increasing access to sensor information gained through a wide variety of platforms, including imagery and navigation satellites, unmanned aerial vehicles (UAVs), and the attack aircraft themselves, which are equipped with advanced sensors and targeting pods. (To assure this access, they need to be in a common network that allows dissemination of sensor information in near-real time.) TACs can use laser range-finders and GPS to determine the coordinates of enemy forces, or they can use laser pointers or designators to show aircraft the enemy's locations. In Afghanistan and more widely in Iraq, U.S. ground forces tracked their positions and disseminated this information to higher headquarters, reducing the risk of fratricide. Map-making has also become more accurate and responsive to the warfighter through electronic distribution. To complete the picture, weapons delivery is becoming more accurate and more reliable. During Operation Anaconda, combat controllers and enlisted terminal attack controllers (ETACs) called in ordnance close to friendly positions without causing fratricide.[2] During operations in

---

[1] Distance safe is the least distance from friendly forces to the point of impact considered acceptable under normal operating conditions.

[2] ETACs are aligned against conventional Army forces, Rangers, and Special Forces. They have great expertise in their specialty and also acquire skills necessary to support particular

Iraq, TACs called in air strikes against Iraqi forces that popped up suddenly within the range of the friendly force's direct-fire weapons.

Jointness Is Descending to Lower Echelons

Currently, and probably well into the future, counterland operations will usually be conducted jointly and within a coalition. Without friendly ground forces to gain information, flush enemy forces, and exploit the destruction of those forces, counterland operations are likely to be ineffective, even against heavily equipped enemy forces, as shown by the largely unsuccessful effort in Kosovo. Moreover, jointness is descending to lower levels of the military hierarchy. Doctrine still reflects processes developed during the Cold War—for example, assuming that the air support operations center (ASOC) will normally be at corps level.[3] But in recent conflicts, arguably even in Iraq, the need for integrated air-land planning existed at division and brigade levels. TACPs are traditionally attached to battalions, but in future contingencies they may be required at company and perhaps even platoon levels to transmit timely information about the battlefield and to fully exploit the emerging capabilities of air power.[4]

Some types of units already practice joint operations at very low levels. SOF lack sustained combat power and rely on air power for their very survival. Air power inserts them, supplies them, keeps them informed, provides them fire, and extracts them when the mission is done. Such forces routinely conduct joint operations at extremely low force levels. In Afghanistan, for example, twelve-man Special Forces

---

Army forces, for example, jump qualification. Combat controllers belong to the SOF community. They are skilled not only in terminal attack control, but also in the operation of austere airfields.

[3] "The ASOC is the primary control agency component of the TACS [Theater Air Control System] for the execution of CAS. Collocated with the senior Army echelon's FSE [fire support element], normally the Corps FSE, the ASOC coordinates and directs fire support for Army or joint force land component operations" (Chairman, Joint Chiefs of Staff, 2003c, p. II-7).

[4] Chapter Six explores this issue in greater depth.

detachments teamed with TACs to support opposition forces, primarily by coordinating CAS. Marine Expeditionary Units include rotary-wing and fixed-wing aviation, attaining capabilities at battalion level that would require joint integration with forces outside the Corps. The Army and the Air Force will also have to operate jointly at lower levels, because Army ground forces are operating in smaller increments. During combat operations in Afghanistan, the largest conventional Army formation was a brigade with three small battalion task forces. During recent combat operations in Iraq, brigade task forces did the fighting and needed to closely integrate their operations with air attacks, not only to prevent friendly-fire incidents, but also to minimize civilian casualties while achieving the desired effects against enemy forces.

Urban combat demands that jointness descend to lower echelons. Urban terrain dissects the battlefield into small segments, often limited to a few blocks and street intersections, where commanders require small combined-arms teams to accomplish their missions. These teams typically include tanks, assault guns, engineer assets, and dismounted infantry supported by indirect-fire weapons and air power. The teams commonly have very restricted views of the terrain, perhaps only to the next block, and they need help quickly to reduce pockets of resistance or to counter enemy attempts to reinforce. Aerial assets can orbit urban terrain, contribute to situational awareness, and attack quickly, as the situation may demand. During urban combat, attack aircraft, fighters, and bombers deliver precise fires that minimize risk to the civilian inhabitants. During recent operations in Iraq, for example, GPS-guided munitions were the weapons of choice in cities because they were more accurate and caused less collateral damage than artillery and rockets did.

Currently, the Army is fielding medium-weight Stryker brigades that will deploy rapidly and operate in nonlinear fashion, implying an increased need for air power. Looking to the future, the Army is developing a new family of manned and robotic combat vehicles that will have minimal passive protection and will operate in highly dispersed fashion. These forces will rely heavily upon their situational awareness to prevail and will be heavily dependent upon air power for deployment, navigation aids, supply, and precise, timely attacks on enemy forces. Typically, these new Army forces will operate as Units of Action that approximate the size of today's brigades. Integration of air and ground forces will need to occur at this echelon.

Kosovo (Operation Allied Force)

Operation Allied Force ended successfully with Serbia's capitulation, but air power was ineffective against the target the NATO allies and the Supreme Allied Commander, Europe, Gen. Wesley K. Clark, most wanted to destroy: the Serb forces conducting "ethnic cleansing" in Kosovo.

Strategy

NATO, led by the United States, conducted Operation Allied Force to stop Yugoslavia's brutal repression of Kosovar Albanians. Like much that has happened recently in the Balkans, the roots of this crisis reach back into the earliest history of the region. Most Serbs regard Kosovo as a heartland of their people, despite its Albanian majority. On June 28, 1389, King Lazar of Serbia suffered a crushing defeat by a Turkish army under Murad I at Kosovo Polje in modern Kosovo. Little is actually known about the battle, but epic poems composed long afterward celebrate it as a pivotal and tragic event in Serb history. On June 30, 1989, the 600th anniversary of the battle, Slobodan Milosevic began his career as a post-Communist politician with a speech near Kosovo Polje. Two years later, Yugoslavia began to disintegrate along national and religious lines, eventually resolving to just Serbia and Montenegro. During the protracted conflict in Bosnia, Kosovo remained relatively peaceful, and the Dayton Accords did not address its status.

During 1997, tension began to mount between Serbs and Albanians in Kosovo. Yugoslavian authorities used harshly repressive measures against a resistance that became increasingly radical. An Albanian organization calling itself the Kosovo Liberation Army (KLA) began conducting paramilitary operations that Yugoslavia denounced as terrorism. The Security Council of the United Nations and the North Atlantic Council blamed Yugoslavia for the escalating violence. In August 1998, NATO Secretary General Javier Solana publicly warned President Milosevic that NATO was preparing military options. Confronted with the prospect of air strikes, Milosevic agreed to allow an unarmed Kosovo Verification Mission to oversee conditions. In mid-January 1999, the group discovered evidence of a massacre of Albanian men by Serb security forces near the village of Racak, prompting the North Atlantic Council to renew its threat of air strikes. In February, Serb and Kosovar Albanian delegates met in Rambouillet, outside Paris, to consider a proposed agreement that would require Yugoslav forces to depart Kosovo and NATO forces to enter. The conference adjourned without result, but the Albanian delegation was later convinced to sign the proposed agreement. After a last visit to Belgrade by U.S. envoy Richard Holbrooke produced no result, the North Atlantic Council decided to initiate air strikes to coerce Yugoslavia.

On March 23, 1999, the day prior to air strikes, Secretary General Solana stated that he had directed Gen. Clark to initiate air operations because Yugoslavia had refused the international community's demands for an interim settlement at Rambouillet. Regarding this action, Solana stated, "It will be directed towards disrupting the violent attacks being committed by the Serb Army and Special Police Forces and weakening their ability to cause further humanitarian catastrophe." On the following day, President Clinton announced three goals: to demonstrate the seriousness of NATO's purpose, to deter attacks on innocent civilians in Kosovo, and to "seriously damage the Serb military's capacity to harm the people of Kosovo."[5] At NATO's Washington summit conference in April, the heads of state affirmed the goals set by the North Atlantic Council on April 12: ending violence and repression in Kosovo, withdrawal of Yugoslav forces, stationing in Kosovo of an international military presence, safe return of refugees, and establishment of a political framework on the basis of

[5] "Statement by the President to the Nation," White House, Office of the Press Secretary, Washington, DC, March 24, 1999.

the Rambouillet agreement. They announced, "We are intensifying NATO's military action to increase pressure on Belgrade."[6] In the following days, NATO cautiously escalated attacks on the Serb infrastructure, especially power generation, but such attacks remained contentious within the alliance.

In response to the NATO air strikes, Milosevic ordered the "ethnic cleansing" of Kosovo, i.e., the expulsion of Albanians. Expulsion began during the first week of the air strikes and continued for two months. By the end of May 1999, 863,000 Kosovar Albanians had been expelled, and hundreds of thousands were displaced internally.[7] Yugoslav forces massacred male inhabitants in areas of suspected KLA activity. The KLA was wholly unable to oppose the expulsion and generally sank to insignificance. By ordering "ethnic cleansing," Milosevic apparently expected to present NATO with a *fait accompli*, but his decision proved to be a fatal strategic blunder. This enormous crime proved that Milosevic wanted the land, not the people, and that his regime could never be trusted to observe human rights insofar as Kosovar Albanians were concerned. It compelled the NATO leaders to persevere, as evidenced by their declaration during the Washington summit. Air power was unable to stop "ethnic cleansing" in Kosovo, an outcome that came as no surprise to senior airmen. Lt. Gen. Michael Short, the air component commander, said, "We couldn't stop the killing in Kosovo from the air. . . . We were not going to be efficient or effective." Gen. John Jumper, commanding U.S. Air Forces in Europe, said, "No airman ever promised that airpower [alone] would stop the genocide that was already ongoing by the time we were allowed to start this campaign."[8]

According to Clark's account, "By mid-May we had gone about as far as possible with the air strikes. The next strategic targets would entail more risk of civilian casualties, and the French, among others,

[6] "Statement on Kosovo," issued by the Heads of State and Government participating in the meeting of the North Atlantic Council in Washington, DC, April 23–24, 1999.

[7] Organization for Security and Cooperation in Europe, 1999. Internal displacement can be estimated only grossly because no international observers were present.

[8] United States Air Forces in Europe, Studies and Analysis Directorate, 2002, p. 19.

were resistant [to further strikes]."[9] To break this impasse, Clark pressed for employment of AH-64 attack helicopters and the Army Tactical Missile System[10] and, if all else failed, for an invasion using land forces. But just at this time, to NATO's surprise and relief, Milosevic capitulated. On June 3, he agreed to accept terms proposed by Finnish President Martti Ahtisaari and Russian envoy Viktor Chernomyrdin, which essentially reiterated NATO's demands. Milosevic had several reasons for capitulating. He realized that he could expect no aid from Russia, and he expected unconstrained bombing if he rejected NATO's terms. In addition, he probably worried about the threat of an invasion.[11] He was little influenced by NATO's ineffective attacks on Serb forces in Kosovo, the target that NATO had given first priority.

Operations

NATO planned Operation Allied Force to last only a few days, on the presumption that Milosevic would capitulate quickly. Initially, the operation was to use only those assets that were immediately available in theater. When Milosevic proved obdurate, the operation gradually expanded in scope and intensity. At the outset, NATO generated fewer than 100 attack sorties per day, but by the end of the effort, it could have generated about 1,000 attack sorties per day had sufficient targets been available.

The area of operations was distant from NATO airbases. For example, Pristina, in eastern Kosovo, lies almost 500 miles from Aviano Airbase in northern Italy. Because of the distance, as more forces arrived in theater, Operation Allied Force required extensive tanker support. By early June, 175 tankers were operating from 12 locations to support operations. Almost 7,000 tanker sorties were flown, not counting the air bridge from North America.[12] Long distances to the

---

[9] Clark, 2001, p. 305.

[10] See Narduli et al., 2002; Clark, 2001, pp. 291, 303–305, 320–321, 331–333, 336–367.

[11] Hosmer, 2001a, pp. xvii, 65–76, 91–107.

[12] United States Air Forces in Europe, Studies and Analysis Directorate, 2002, p. 27.

area of operations necessarily reduced times on station, even with the help of aerial refueling.

Lack of targeting data was the single greatest constraint on air operations. Targeting within Serbia was limited by the reluctance of NATO partners to inflict suffering on the civilian population. NATO gradually and reluctantly expanded the target sets in Serbia to include the electrical power grid, bridges over the Danube River, petroleum refineries, and communications. By late May, NATO had almost exhausted the infrastructure targets it was willing to strike, but fortunately, Milosevic seems not to have grasped this circumstance. Targeting within Kosovo was limited by NATO's inability to detect and identify Serb forces that were dispersed and concealed in forests and villages.

Although NATO quickly achieved air superiority over the entire airspace of the former Yugoslavia (Kosovo, Montenegro, and Serbia), the surviving air defense, including mobile SA-6 batteries, antiaircraft artillery, and man-portable air-defense missiles, had important effects. It compelled NATO to continue suppression efforts indefinitely, requiring nearly constant support from F-16CJ and EA-6B aircraft, supplemented by German and Italian Tornados. It practically eliminated operations by AC-130 gunships, which with their advanced sensors and firepower might otherwise have been the most effective aircraft against fielded Serb forces. The low-level threat convinced U.S. decisionmakers to reject Clark's plan to employ Army AH-64 attack helicopters, organized as Task Force Hawk, against Serb forces in Kosovo. The threat to these forces and the risk of collateral damage outweighed potential gains unless the rules of engagement were relaxed to permit heavy suppressive fires. But such fires would have imperiled civilians and might have discredited the entire air operation.[13] The residual air defense, especially the threat from man-portable air-defense missiles, also impelled NATO to keep its attack aircraft at medium altitude, generally above 15,000 feet. However, even had

---

[13] For a discussion of Task Force Hawk, see Nardulli et al., 2002, pp. 80–97.

aircraft routinely flown below cloud cover, they would still have had difficulty finding Serb forces, which were usually well hidden from aerial observation.

Operation Allied Force saw important innovations, including the first employment of the joint direct-attack munition (JDAM) and the Predator UAV in combat. JDAMs are general-purpose bombs mated to kits that use inertial navigation and GPS for guidance. After the operator has entered coordinates and released the JDAM within its computed envelope, the weapon guides itself autonomously, using space-based geopositioning, to the target coordinates set at the time of release. On March 24, 1999, B-2 aircraft first attacked targets in Serbia with the GBU-31, a JDAM variant mating a guidance kit to a 2,000-pound general-purpose bomb. Using onboard radar, the B-2 crews were able to image the targets and refine the initial sets of coordinates to improve accuracy.[14] With aerial refueling, B-2 aircraft were able to fly from Whiteman Air Force Base, Missouri, to their targets without requiring a forward base. While initially used against fixed targets, JDAM was suitable for any targets whose coordinates were known. In addition to being relatively inexpensive, yet precise, JDAMs are insensitive to weather. They can be dropped from high altitude without degrading accuracy. Their guidance penetrates cloud cover, so their performance is generally unaffected by weather, as long as the coordinates can be obtained. By contrast, laser-guided bombs are highly sensitive to weather, which attenuates or blocks sighting for the weapon and the guiding beam.

During Operation Allied Force, the Predator was flown from Tuzla, in Bosnia, into Kosovo airspace, usually to collect data against Serb ground forces. Predator could loiter quietly for hours in enemy airspace at low altitudes without risking a pilot's life while sending near-real-time data to remote locations. These data included high-quality streams of electro-optical and infrared images and, in later models, radar returns. Given the stringent rules of engagement,

---

[14] Lambeth, 2001, p. 91.

Predator was especially useful for positive identification of targets glimpsed by other means. Some Predators were equipped with laser designators, but this capability remained unused. During the course of the operation, Predators were lost for various reasons, including weather conditions and navigational error. The Army flew Hunter UAVs from Skopje in Macedonia.

Insights

In official documents, counterland attacks in Kosovo were termed "close air support," although there were no friendly land forces to support. The term was chosen because the procedures, including positive target identification, most closely resembled those normally used for CAS. To ensure positive identification as required by the rules of engagement, the procedures for counterland in Kosovo resembled those normally used for CAS, except that no ground forward air control was available. Airborne forward air control was severely restricted by the air-defense threat and persistent bad weather. Attack pilots might receive final instructions from the Combined Air Operations Center while they completed aerial refueling outside Kosovo airspace. As they entered this airspace, they would normally contact an airborne control center in the EC-130E/J aircraft, whose controllers would brief them on the situation and hand them to FACs flying A-10 and F-16 aircraft. These FACs would talk the attack pilots onto their targets, taking care to assure that the pilots had correctly identified the targets before being cleared to strike. The pilots could engage with AGM-65 Mavericks, GBU-12 laser-guided bombs, CBU-87 cluster bombs, and general-purpose gravity bombs. Maverick and laser-guided bombs were preferred against pinpoint targets, such as combat vehicles, but these could seldom be identified. CBU-87s had devastating effects against area targets, such as vehicle convoys, but they had to be employed with extreme care to avoid causing collateral damage and littering the country with unexploded submunitions. In deference to its NATO allies, the United States refrained from employing the CBU-89, which scatters a mix of antitank and antipersonnel mines.[15]

To avoid low-level air defenses, attack aircraft were initially kept at medium altitude, above 15,000 feet. At this altitude, airborne observers had difficulty distinguishing between military and civilian vehicles, especially during bad weather. This difficulty caused the gravest incident of collateral damage during Operation Allied Force, an inadvertent attack on refugees on the road between Djakovica and Decane during the afternoon of April 14, 1999. The attack began when an F-16 pilot observed a small group of vehicles near villages that were in flames, presumably ignited by Serb forces that were conducting "ethnic cleansing." Jaguar and F-16 pilots attacked one or more groups of vehicles in the area, using GBU-12 bombs, but cockpit video subsequently revealed that civilian vehicles were present.[16] Human Rights Watch documented 73 civilians killed and 36 wounded in the Djakovica incident.[17] Lt. Gen. Michael Short, serving as the air-component commander, subsequently authorized flight at lower altitudes for FACs and for pilots during their final approaches to targets, but target identification remained difficult.

The unsolved tactical problem in Kosovo was how to identify Serb ground forces with enough certitude to preclude incidents like Djakovica. Bad weather and heavy vegetation exacerbated this problem but did not cause it. The cause was that the Serbs gave hiding from air attack their highest priority. Serb forces were not confronted with any challenge on land that compelled them to deploy in arrays that could be targeted easily. Serb commanders were concerned about NATO deployments to Macedonia, but not to the point of concentrating their forces on the attack corridors. The KLA did not have

---

[15] The NATO allies are signatories to the Ottawa Treaty (1997), by which the parties agree never to use antipersonnel mines under any circumstances. The United States is not a signatory.

[16] North Atlantic Treaty Organization, 1999; Dobbs and Vick, 1999; Priest, 1999; *Final Report to the Prosecutor*, 2000.

[17] Arkin, http://hrw.org/reports/2000/nato/Natbm200-01.htm, pp. 11–12. Despite this incident, there were relatively few civilian deaths during Operation Allied Force: About 500 civilians were killed in 90 incidents, according to Arkin's investigation.

enough combat power to make the Serbs array large forces against it. Driving Kosovar Albanians from their homes, so-called "ethnic cleansing," required only small forces that were intermingled with their victims and hence extremely difficult to target. Indeed, the Serb response to Operation Allied Force was to greatly accelerate the "ethnic cleansing."

Despite months of effort, air attacks did little damage to Serb forces. Those in Kosovo not only survived, they even received reinforcement while Operation Allied Force was in progress. The rules of engagement allowed the Serbs to operate helicopters, which continued to conduct resupply missions. Gen. Clark, the Supreme Allied Commander, Europe, subsequently claimed "successful strikes" against 93 tanks, 153 armored fighting vehicles, and about 389 pieces of artillery and mortars.[18] But an on-site survey by a munitions-effectiveness assessment team confirmed destruction of only 14 tanks, 18 armored fighting vehicles, and 20 pieces of artillery.[19] Even in the Mount Pastrik area, where B-52s had attacked Serb forces engaged with the KLA, the team found no wreckage of military equipment, despite an intensive search.

Afghanistan (Operation Enduring Freedom)

Air power enabled the indigenous Afghani opposition, supported by a few SOF, to sweep the Taliban from power with dramatic suddenness but was far less effective against Taliban remnants and al Qaeda after they ceased to operate in conventional fashion. An engagement in the Shah-i Kot Valley exposed difficulties in air-land coordination.

Strategy

The U.S. conducted Operation Enduring Freedom to eliminate al Qaeda following the September 11, 2001, attacks that destroyed the

---

[18] Clark and Corley, 1999.

[19] Barry and Thomas, 2000; Letters to the Editor, *Air Force Magazine*, August 2000, pp. 6–7; Lambeth, 2001, pp. 131–132.

World Trade Center and damaged the Pentagon, killing several thousand people. The timing and magnitude of these attacks were surprises, but al Qaeda's intent to attack the United States was well known. Osama bin Laden, the son of an immensely wealthy construction magnate, founded al Qaeda (which means "the base") in the late 1980s to oppose the Soviet occupation of Afghanistan. In 1992, bin Laden established legitimate businesses in the Sudan as a cover for terrorist activities. Under pressure from the United States and Saudi Arabia, Sudan expelled bin Laden in 1996. He returned to Afghanistan, where the fundamentalist Taliban regime gave him sanctuary. On February 23, 1998, bin Laden and several other radical leaders issued a statement, in the preamble of which they denounced U.S. "occupation" of the Arabian Peninsula and U.S. "aggression" against Iraq. They then issued a *fatwa* (religious decree) that declared: "The ruling to kill the Americans and their allies, civilians and military, is an individual duty for every Muslim who can do it in any country in which it is possible to do it." On August 7, 1998, al Qaeda detonated explosives at the U.S. embassies in Kenya and Tanzania, killing 224 people, including 12 U.S. nationals. In response, President Clinton ordered cruise-missile attacks on terrorist camps in Afghanistan and a pharmaceutical plant in Khartoum that was believed to be engaged in production of chemical weapons for al Qaeda. In October 2000, al Qaeda attacked the USS *Cole*, an Arleigh Burke–class destroyer that had stopped to take on fuel at Aden. A small boat, which had helped moor the *Cole*, exploded against the port side, tearing a large hole in the hull and killing 11 sailors.

On September 11, following the attacks on the Pentagon and the World Trade Center, President Bush declared, "We will make no distinction between the terrorists who committed these acts and those who harbor them."[20] Two days later, in a meeting of the National Security Council, Director of Central Intelligence George Tenet briefed President Bush on a plan to invigorate the Northern Alliance, a loose alliance of mainly Uzbek and Tajik groups locked in a stalemated war

[20] "Statement by the President to the Nation," White House, Office of the Press Secretary, Washington, DC, September 11, 2001.

against the Taliban, which was dominated by Pashtuns. Unfortunately, al Qaeda operatives posing as journalists had already assassinated the most able opposition leader, Mohammed Shah Masood. On September 15, in a meeting at Camp David, the Chairman of the Joint Chiefs of Staff, Gen. Hugh Shelton, briefed President Bush on three options for the use of military forces in Afghanistan. Two days later, in Washington, President Bush announced his decision to execute Shelton's third and most ambitious option, which included missile strikes, extensive air attacks, and the insertion of small ground forces.[21] Publicly, the United States presented the Taliban regime with a list of demands, which included closing terrorist camps, handing over the leaders of al Qaeda, and returning foreign nationals unjustly detained in Afghanistan. On October 7, after none of these demands were met, the United States initiated air strikes in Afghanistan. The President said, "Today we focus on Afghanistan, but the battle is broader. Every nation has a choice to make. In this conflict there is no neutral ground. If any government sponsors the outlaws and killers of innocents, they have become outlaws and murderers themselves. And they will take that lonely path at their own peril."[22] Despite having declared the Taliban regime to be an outlaw, the President had not publicly made its demise a goal of U.S. policy. However, in internal meetings, his advisers were already discussing policy toward a post-Taliban Afghanistan.[23]

In contrast to Serbia, the Afghan infrastructure was not considered an appropriate target, with few exceptions. The United States wanted to counter bin Laden's propaganda by demonstrating its friendliness toward the population of Afghanistan, for example, by air-dropping humanitarian daily rations. Attacking infrastructure targets would have contradicted this policy. Moreover, damage caused during military operations would make reconstruction of a post-

[21] Woodward, 2002, pp. 74–98.

[22] "Presidential Address to the Nation," The Treaty Room, Washington, DC, October 7, 2001.

[23] Woodward, 2002, pp. 192–193, 195.

Taliban Afghanistan that much more difficult. Finally, Afghanistan was so impoverished as to have little infrastructure worth attacking, even if the United States had been so inclined. As a result, the United States concentrated on the Taliban's fielded forces.

Beginning in mid-October, the United States deployed small teams composed of Special Forces detachments and Special Tactics combat controllers to support the Northern Alliance and other opposition forces in Afghanistan. The key function of these teams was to call in air attacks on Taliban forces in contact with opposition forces. Initially, U.S. planners had little confidence in the Northern Alliance's ability to exploit the effects of these air attacks. The U.S. Central Command had a plan to introduce about 50,000 U.S. ground forces if the Northern Alliance failed to make progress.[24] Such a Herculean effort proved unnecessary when the Northern Alliance, with SOF, had spectacular success, seizing Mazar-e Sharif on November 10 and Kabul two days later. These successes demoralized the Taliban and enabled small Pashtun opposition forces heavily supported by air power to seize Kandahar on December 7, effectively ending Taliban rule. But this dramatic success was merely a precondition to achieving the fundamental U.S. goal of eliminating a safe haven for terrorists in Afghanistan.

Al Qaeda forces proved far more elusive than the Taliban, and the U.S. achieved less-decisive results against them. The formula for success against the Taliban, i.e., indigenous forces supported by air power, produced disappointing results against al Qaeda. Starting in the first days of Operation Enduring Freedom, the United States attacked targets in the White Mountains, where al Qaeda was known to have key facilities. During December, U.S. Central Command made a major effort against al Qaeda in the Tora Bora area of the White Mountains, but indigenous forces performed poorly and Osama bin Laden escaped, apparently to Pakistan. In March 2002, a newly created U.S. task force initiated Operation Anaconda against al Qaeda and Taliban remnants in the Shah-i Kot area. Operation Ana-

---

[24] Woodward, 2002, pp. 291–292.

conda encountered unexpectedly strong opposition, which was overcome by air strikes, the heaviest of the war. Thereafter, the enemy reverted to sporadic hit-and-run attacks on U.S. forces. The war in Afghanistan ended with incomplete success. The U.S. had swept away the Taliban regime, denying al Qaeda its sanctuary, but remnants of the terrorist organization survived, especially in tribal regions along the Afghanistan-Pakistan border.

Operations

Basing constraints severely affected air operations in Afghanistan. Long-standing tension with Iran precluded basing U.S. aircraft in that country. With opinion in Pakistan heavily opposed to U.S. intervention against the Taliban, the Pakistani government refused to allow basing of attack aircraft. Uzbekistan allowed basing, but its former Soviet airbases were woefully inadequate to support sustained operations. As a result of these constraints, most attack sorties were flown either by fighters based on carriers in the Arabian Sea or by bombers flying from Diego Garcia. Even with three or four aerial refuelings, F-14 and F/A-18 aircraft had short times on station in Afghanistan, often as little as 15 minutes. The F-14s and F/A-18s often carried two GBU-12s, while the F-16s usually carried four and the F-15Es carried nine. B-1 and B-52 aircraft could remain on station for several hours at a time and therefore constituted the most reliable source of CAS. But their precision-guided munitions were limited to JDAMs, which were difficult to employ against pinpoint and moving targets.

As noted above, the campaign in Afghanistan initially centered on special operations to support Northern Alliance forces against the Taliban. The Northern Alliance was not a homogeneous organization, but a loose confederation of several groups opposed to Taliban rule. Initially, the most active and important groups were those controlled by Mohammed Qasim Fahim, Rashid Dostum, and Mohammed Attah. Fahim was a Tajik from the Panjshir Valley, who had inherited command over forces in the Bagram area from the great Tajik leader Ahmed Shah Massoud, following his assassination by al Qaeda. Dostum, an Uzbek with a well-deserved reputation for ruthlessness, was opposing Taliban forces south of Mazar-e Sharif. Attah was a Tajik from northern Afghanistan and Dostum's rival within the Northern Alliance.

Starting on September 26, the Central Intelligence Agency (CIA) began infiltrating small teams into Afghanistan to contact leaders of the Northern Alliance.[25] These teams brought large amounts of currency to finance operations against the Taliban and prepared for the introduction of U.S. forces. On October 19, Task Force Dagger, later designated Joint Special Operations Task Force–North, began infiltrating teams of Army Special Forces soldiers and Air Force combat controllers by MH-47s into Afghanistan. Each team consisted of an Operational Detachment Alpha, normally 12 men, and one or two combat controllers from the 720th Special Tactics Group. Conventional TACs later supplemented the Air Force special operators. On October 19, the first two teams arrived: Tiger 1, assigned to Fahim in the Panjshir Valley and Bagram, and Tiger 2, assigned to Dostum, then about 50 miles south of Mazar-e Sharif. On November 2, Tiger 4 arrived to support Atta, operating to the east of Dostum.

Dostum proved to be the most aggressive of the Northern Alliance leaders. He was extremely hospitable to Tiger 2, which he obviously expected would tip the balance in his favor. Indeed, he was so solicitous of the team's safety that he tended to keep its members too far removed from the fighting.[26] Dostum's forces consisted of a few hundred horsemen riding with him, who might be reinforced by several thousand militiamen in a particular battle. They were equipped with light-infantry weapons, in contrast to the Taliban, which had tanks, air-defense cannons, and artillery left over from the Soviet occupation. Air power almost immediately made a decisive difference. It devastated the entrenched Taliban forces and demoralized them, while heartening the opposition fighters. After about three weeks of hard fighting, Dostum and Attah entered Mazar-e Sharif on November 9. When this northern city fell, most of the surviving Taliban and

---

[25] Woodward, 2002, pp. 139–148.

[26] Moore, 2003, pp. 66–67.

al Qaeda forces fled toward Konduz. Fahim's forces prevented their escape to the south, and they were eventually compelled to surrender.

Fahim's forces, led in the field by Bismullah Khan, were generally deployed opposite the Taliban deployed on the Shomali Plain in defense of Kabul. The line of confrontation ran through the old Soviet airbase at Bagram, about 30 miles north of the capital. From the control tower at the airbase, combat controllers had excellent visual observation of the Taliban positions, which extended for several thousand yards to their front. After several postponements due in part to poor weather, Tiger 1 finally began calling in air strikes on October 20. For several weeks, Air Force and Navy aircraft attacked trenches, bunkers, vehicles, and command posts, while Taliban forces replied with inaccurate and desultory artillery fire. Starting on November 10, the Taliban defenses were subjected to increasingly heavy air strikes, and two days later, Fahim's forces started a general advance, quickly overrunning the Taliban positions and taking Kabul on November 13. Many Afghani adherents of the Taliban changed sides, but hard-core Taliban and al Qaeda members fled south.

In contrast to northern Afghanistan, there was little opposition to Taliban rule in the south, making operations there far more problematic. On November 14, Texas 12, a team composed of CIA agents, Army Special Forces soldiers, and one Air Force combat controller, arrived by MH-60 helicopters in the highlands north of Kandahar to support Hamid Karzai, who later became President of Afghanistan. On November 19, a similar team designated Texas 17 was inserted near the Pakistani border southeast of Kandahar to support Gul Agha Shirzai. In contrast to the Northern Alliance, Karzai and Shirzai had only a few hundred ill-equipped fighters to oppose much larger forces. Although defeats in the north had disheartened the Taliban, Texas 12 was challenged by the task of keeping Karzai alive while he negotiated Taliban capitulation. During a battle near Tarin Kowt on November 18, air attacks prevented the Taliban from defeating Karzai's small band of followers.[27] Relentless air support

[27] Moore, 2003, pp. 196–202.

allowed Karzai and Shirzai to advance toward Kandahar from two directions. In a series of negotiations, Karzai won over the defenders of Kandahar, except for a few irreconcilables who fled the city with Mullah Mohammed Omar. On December 9, Karzai and Shirzai entered Kandahar, welcomed by the city's inhabitants, who had chafed under Taliban rule.

In retrospect, Taliban rule was far less stable than it appeared, in large part because its fanaticism alienated the Afghan people. The Taliban had originated in regional seminaries (*madrassas*) as a puritanical religious movement. It spread throughout the Pashtun-inhabited areas of Afghanistan in a spasm of popular enthusiasm, but within a few years, this enthusiasm waned, especially in urban areas where Afghanis had long practiced a more tolerant Islam. Had Taliban rule been more popular, the Pashtun people in southern Afghanistan might well have rallied against the Northern Alliance, which was dominated by Tajik and Uzbek peoples. As it was, defeats in the north inspired most Afghanis to change sides, and they generally did so with impunity. As a foreign presence, the members of al Qaeda could not simply change sides. They had to either go down fighting or flee for their lives when the Taliban was no longer able to protect them.

Once Taliban rule had collapsed, the United States turned its full attention to remnants of the Taliban and al Qaeda lurking in the extremely rugged terrain of the White Mountains. Once again, the United States tried to exert leverage with indigenous forces—in this case, fighters loyal to Hazart Ali, the governor of a province in southeastern Afghanistan. But Ali's followers showed little enthusiasm for a winter expedition against al Qaeda. The Marines, now established at Kandahar Airport, planned to conduct blocking operations but did not receive an order to execute. Pakistan declined to allow U.S. forces on its territory but agreed to block passes leading south from the White Mountains. As a result, the United States had only its own small special operations teams to hunt an elusive enemy that was well acquainted with the ground and had numerous hiding places. Bombing initially concentrated in the Tora Bora area, where bin Laden was thought to be hiding. Later, the United States systematically closed caves by aiming JDAMs above the entrances. For a few weeks, U.S. planners thought that bin Laden might have died in the strikes on Tora Bora. SOF collected tissue samples for comparison with samples taken from his close relatives,[28] but his survival was subsequently confirmed by a series of voice tapes in which he taunted the United States for its failure.[29] Air power had devastated the Taliban's fielded forces but had proven much less effective against small groups of al Qaeda that were trying to hide.[30]

Operation Anaconda was the first large-scale conventional operation involving U.S. forces in Afghanistan. It was controlled by Combined Joint Task Force (CJTF) Mountain with headquarters in Bagram. This headquarters was drawn from the 10th Mountain Division and was not well prepared to handle joint operations. The plan for Operation Anaconda envisioned Afghan forces under Zia Lodin sweeping down the Shah-i Kot Valley while U.S. forces blocked escape eastward toward Pakistan. On the basis of fragmentary intelligence, the U.S. commander expected to encounter about 200 al Qaeda members, who would presumably attempt to flee. As a result, he anticipated little need for CAS. Early on March 2, 2002, Afghan forces began their advance, but they retreated after an AC-130 aircraft inadvertently attacked their lead element. Task Force Rakkasan, organized around the 3rd Brigade, 101st Airborne Division (Air Assault), air assaulted into the valley the same morning and almost immediately came under fire from an enemy armed with small numbers of man-portable air-defense missiles, rocket-propelled grenade launchers, mortars, and heavy machine guns. One small battalion task force from the 10th Mountain Division came under especially heavy fire near the position designated as Ginger, near the southern end of the valley.

The Task Force Rakkasan commander and his air-liaison officer also came under fire as they were exiting a UH-60 on high ground

[28] Rose, 2002.

[29] For a summary of failed attempts to kill Osama bin Laden, see Mayer, 2003.

[30] Barry, 2002; Donnelly, 2002; Forney, 2001; Smucker, 2002.

overlooking the valley. During the first day, the air-liaison officer gave highest priority to the battalion pinned down near Ginger. Unfortunately, he had only modest air assets immediately at his disposal, because the planners had expected little opposition. Task Force Rakkasan called its AH-64 Apache helicopters against enemy positions in the valley, but they could not survive the intense ground fire coming from every direction. Only one Apache was still flying at the end of the day, but fortunately, none of the pilots was seriously injured. Soldiers on the ground suffered no fatalities, although many were wounded, especially in the battalion from the 10th Mountain Division. This battalion was extracted from its untenable position near Ginger during the first evening. The following day, Task Force Rakkasan reinforced its positions in the valley and began a systematic sweep. The overall commander declared villages from which fire was coming to be hostile, and they were attacked by bombers directed from the Combined Air Operations Center in Saudi Arabia. Much larger numbers of fighters and bombers attacked targets throughout the valley and on mountain roads and trails, referred to as "ratlines," extending to the east. After two days, the volume of enemy fire began to fall off sharply. After about three days, resistance on the ground ceased entirely, but air attacks continued unabated; very heavy strikes were made on Takur Ghar, a mountain next to a "ratline" at the southern end of the Shah-i Kot Valley, where SOF had previously fought a desperate engagement.

Early on March 4, SOF using an MH-47E attempted to insert a reconnaissance element on Takur Ghar, also called "Roberts Ridge," just before dawn. A Navy SEAL fell out of the aircraft when it came under heavy fire and climbed steeply. A SEAL team attempted to recover him but was driven away by fire that resulted in the loss of an Air Force combat controller. Due to missed communications, a quick-reaction force of Army Rangers in two MH-47Es was sent to Takur Ghar, not knowing that the ridge was under enemy fire. The first helicopter was downed by fire that killed the right-side gunner and wounded both pilots. One Ranger was killed in the aircraft, and two more were killed as they attempted to exit. By this time, the ridge was in full daylight and the covering AC-130 had been ordered to depart the area due to the threat from man-portable air-defense missiles. Pinned down by fire, the Rangers lacked the strength to clear the enemy forces above them on the ridge. To protect the friendly position, Air Force personnel called in air strikes, first strafing and then bombing runs, within distance safe. The second helicopter discharged its Rangers well below the position, and they climbed laboriously upward. When they reached the position about mid-morning, the combined force successfully cleared the enemy from the higher ground. The enemy began firing from behind the downed helicopter, mortally wounding an Air Force pararescueman who was assisting the wounded. Again a combat controller called in air strikes very close to the friendly forces' position. After darkness fell, the entire force was lifted off Takur Ghar.[31]

Operation Anaconda continued until March 11, although U.S. forces had little contact on the ground after the first three days. Air forces, including A-10 aircraft that had deployed into the region on March 3, attacked enemy forces throughout the area, including those on the "ratlines" leading toward Pakistan. During this phase of the operation, the pilots of A-10 aircraft also functioned as FACs.

Insights

Operation Enduring Freedom was a remarkable success for Army Special Forces and Air Force combat controllers, who supported indigenous forces with air strikes. However, the operation also pointed out shortfalls in processes, organization, training, and equipment. Organizationally, Army Special Forces were aligned with conventional Air Force TACPs, not with Air Force special operators (combat controllers). During peacetime, the 19th Air Support Operations Squadron was aligned with the 5th Special Forces Group for training. In wartime, the 720th Special Tactics Group, not the 19th Air Support Operations Squadron, provided combat controllers. This organization assured that the 5th Special Forces Group received highly qualified operators, but not operators with whom they had trained.

---

[31] Department of Defense, 2002; Graham, 2002a,b.

Expectations and training standards for terminal attack control processes were contentious issues for some units. Army Special Forces tended to think that calling in air attacks was analogous to calling in artillery, a skill that any competent infantry soldier could master. They were trained to a lower standard than controllers were, but they were familiar with emergency CAS using the nine-line format. During combat, they were outraged when Air Force and Navy pilots reacted skeptically to calls that seemed amateurish or irregular. Pilots and aircrew understandably expected a high degree of precision and clarity. The enormous risk of even small mistakes was illustrated by an incident on December 5 in which several members of Texas 12, the team supporting Hamid Karzai, were killed. The combat controller had been on duty continuously for a long period of time. He was relieved by an ETAC who was not completely familiar with the combat controller's equipment. During an attack by Taliban forces, he changed batteries in an unfamiliar GPS receiver without realizing that when repowered, it would revert to its own location rather than the recently lased target. As a consequence, he inadvertently directed an orbiting B-52H aircraft to attack his own location. The attack with GBU-31s killed three Special Forces soldiers and several Afghani fighters and seriously wounded a combat controller. It might easily have killed Karzai, who learned minutes after the attack that he had been appointed to lead the new provisional government of Afghanistan. This incident could have been prevented by more thorough training or by having equipment with fail-safe features, such as an internal feature that would tag the operator's own location.

Controllers' equipment has been procured as separate items, with commercial products filling some niches. As a result, controllers in Afghanistan had to employ several different pieces of equipment, including spotting scopes, GPS receivers, laser range-finders, and laser designators, each with its individual power requirements, rather than an integrated system. The controllers usually lacked the ability to obtain the exact coordinates needed for the new JDAMs, and they also lacked the ability to transmit these coordinates automatically machine-to-machine. Every manual transfer of data opened a new possibility for error. Angular errors in sighting and errors in calculating elevation, compounded with the weapons' inherent error probability, generated frustrating inaccuracies. Without insight into what caused these errors, controllers were hard pressed to attain the precision potentially available with the new munitions. Moreover, controllers had to transmit and check coordinates by voice communication, which functioned only when the aircraft was overhead and in their line of sight. They could have saved time and avoided potential errors in transmission if their equipment had provided a direct data link to the attack aircraft.

The MQ-1 Predator armed with two Hellfire missiles was employed for the first time in combat during Operation Enduring Freedom. It loitered at relatively low altitude over target areas without risk to a pilot while providing fairly high-quality streaming video and a modest attack capability. It transmitted reconnaissance data to various command posts and higher headquarters and also to at least one aerial platform, the AC-130 gunship. It became extremely useful against time-sensitive targets, such as Taliban and al Qaeda leadership. When available, it gave excellent coverage of unit-level engagements, for example, the engagement at Roberts Ridge. This combat debut suggested the immense potential for UAVs over the battlefield, but it also revealed some pitfalls. One was the tendency of higher headquarters staffs to focus attention on events within Predator's very narrow field of vision because of their fascination with the video, thereby affecting priorities.

The AC-130 gunship provided invaluable support to special operations and was also employed successfully with conventional forces, for example, in the Shah-i Kot Valley. More than any other platform, it combined loiter time, multiple sensors, precise fire, and situational awareness of friendly and enemy forces. It would have been an ideal weapon for operations in Afghanistan had it been less vulnerable to air defense. Due to the limited ranges of its weapons, the AC-130 had to orbit at altitudes where it was vulnerable to attack by man-portable surface-to-air missiles. As a result, the AC-130 was generally prohibited from flying daylight missions.

The Air Force is currently studying alternatives for a next-generation gunship that would be much less vulnerable to enemy fire.

The next-generation gunship would provide long loiter time, continuous surveillance and tracking, and rapid engagement with precise weapons as the AC-130 does, while being much less vulnerable. To become less vulnerable, the next-generation gunship would probably need standoff weapons, allowing it to stay beyond range of anti-aircraft artillery and man-portable missiles.

Finally, and most importantly, experience during Operation Anaconda and its aftermath highlighted some shortfalls in joint control of Army and Air Force forces deployed in the area. Maj. Gen. Franklin L. Hagenbeck, commanding the 10th Mountain Division, was directed on short notice to create a task force for all conventional forces in Afghanistan (CJTF Mountain) without being given an appropriate joint staff. His headquarters in Bagram was derived primarily from his divisional headquarters and lacked an ASOC, normally found at corps level, or a full-time liaison with the Combined Air Operations Center. It also appears that Hagenbeck's staff failed to coordinate adequately with the Combined Air Operations Center during the planning process.

In an interview after the operation, Hagenbeck said that the Air Force refused to allow personnel other than its own controllers to call in precision munitions, yet it failed to provide enough TACs for Army units. As a solution, he recommended training "universal observers," including personnel drawn from artillery units.[32] In fact, there were enough TACs in the Shah-i Kot Valley, perhaps even too many for the constrained battle space, e.g., several controllers were calling for missions against the same target. There might have been more controllers had the battalion from the 10th Mountain Division not left its aligned TACP behind when it deployed earlier to Karshi Khanabad. Although he was incorrect in the specific case, Hagenbeck may still have identified a future requirement. As the Army trans-

---

[32] Hagenbeck, 2002. If the designation of "universal observer" implies being qualified to call for fire, identify and geolocate targets, adjust aimpoints, and assess battle damage, the study team envisions implementing this concept in the far term. See "Disaggregate the Terminal Attack Function" in Chapter Six. If a "universal observer" implies being qualified to perform all the functions of a forward observer and a TAC, we would not recommend implementing the concept without further study.

forms, it will increasingly need CAS, and it will need to have the terminal attack function performed at lower echelons than in the past. Air Force Chief of Staff Gen. John P. Jumper responded to Hagenbeck's criticism by initiating high-level contacts between the Air Force and the Army.[33]

Response times were a problem for engaged ground forces during Operation Anaconda both for technical reasons and because of the rules of engagement in effect at the time. Many of the air attacks were intended to suppress enemy fires, making quick response a necessity. But a bomber employing JDAM usually required about ten minutes between drops to enter new coordinates and to reposition itself, a considerable time when friendly troops were under fire.[34] Using several aircraft could reduce the time between attacks, but it would have added greatly to the complexity of the operation. The rules of engagement allowed free use of weapons to support friendly troops in contact with the enemy but were more restrictive in other circumstances, especially when there was risk of collateral damage. For example, attacking vehicles on the "ratlines" required positive identification, i.e., substantial proof that the vehicles were associated with al Qaeda and were not innocent traffic. Despite all their difficulties, including those attributable to inadequate planning, fixed-wing aircraft were the primary killers during Operation Anaconda. They compensated for Rakkasan's lack of organic firepower by delivering heavy concentrations of ordnance on targets all over the valley without causing friendly casualties, except for one inadvertent attack by an AC-130 aircraft on a column of friendly Afghan forces. During the critical first day, AH-64 attack helicopters could not provide effective CAS due to intense ground fire in the narrow Shah-i Kot Valley.[35]

---

[33] Grossman, 2003.

[34] Interviews with personnel from the 93rd Bomber Squadron Air Force Reserve at Barksdale Air Force Base, Louisiana, July 24, 2003, and with Capt. Paul "Dino" Murray, Air Liaison Officer, 1st Brigade, 101st Airborne Division, in RAND's Washington, DC, office, September 27, 2002.

[35] See Lambeth, forthcoming.

Iraq (Operation Iraqi Freedom)

In Operation Iraqi Freedom, the top Iraqi leadership survived repeated air strikes, but air power was overwhelmingly effective against Iraqi land forces. It helped render the regular Army and Republican Guard almost completely ineffective. It was devastating when employed against Iraqi paramilitary forces, even at very close distances to friendly forces. Air-delivered munitions were often weapons of choice in urban areas because they penetrated hard structures reliably and precisely, with less risk of collateral damage than indirect-fire weapons presented. As in Afghanistan, TACs were the vital link between air and land forces.

Strategy

The U.S. conducted Operation Iraqi Freedom to disarm Iraq of weapons of mass destruction (WMD), end Iraqi support for terrorism, and free the Iraqi people from Saddam Hussein's regime. This operation was the culmination of a protracted confrontation with Iraq that began on August 2, 1990, when Iraq invaded Kuwait in an attempt to seize its oil wealth. In response to the invasion, the United States led a coalition effort to liberate Kuwait, under a mandate from the Security Council of the United Nations.

Operation Desert Storm began on January 17, 1991, with air attacks that continued until February 24. During this phase, coalition air forces struck a wide variety of targets throughout Iraq and Kuwait, including air defense installations, ballistic missiles, WMD, leadership, communications, transportation, and ground forces. After the first few weeks, Gen. H. Norman Schwarzkopf, commanding the U.S. Central Command, shifted the weight of air attack to Iraqi ground forces deployed against the coalition. Using a mixture of imagery analysis and subjective evaluation, Schwarzkopf estimated that all Iraqi units on the front line had been bombed down to 50 percent strength or less before ground operations began.[36] Ground operations

[36] Schwarzkopf, 1992, pp. 431–432, 439.

lasted only four days against generally weak and uncoordinated resistance. Operation Desert Storm was a classic demonstration of air-ground synergy. Coalition ground forces fixed Iraqi ground forces, making them good targets for air attack. Destroying and demoralizing Iraqi ground forces until they became ineffective, coalition ground forces gained a rapid victory.

In the aftermath of his military defeat, Saddam Hussein agreed to accept UN inspectors to verify disarmament, i.e., removal of Iraq's WMD, but he consistently obstructed their efforts. In its attempt to contain and coerce Iraq, the United States enforced no-fly zones encompassing much of Iraqi airspace. During a decade-long effort, Operation Northern Watch, based in Turkey, flew more than 16,000 sorties, and Operation Southern Watch, based primarily in Saudi Arabia and Kuwait, flew more than 200,000 sorties.[37] During the same period of time, the United States led a number of operations directed against Iraq. In October 1994, U.S. Central Command conducted Operation Vigilant Warrior, a deployment of forces to the Persian Gulf region in response to a threat of Iraqi aggression against Kuwait. In September 1996, it conducted Operation Desert Strike, a series of cruise-missile strikes in response to Iraqi repression of Kurds and Shi'ite Muslims. In December 1998, it conducted Operation Desert Fox, a four-day air operation against Iraqi WMD, command and control, and selected facilities associated with the Republican Guards.

In his 2002 State of the Union Address, President George W. Bush identified North Korea, Iran, and Iraq as regimes that supported terrorism. He said, "I will not stand by, as peril draws closer and closer. The United States of America will not permit the world's most dangerous regimes to threaten us with the world's most destruc-

---

[37] Formal Witness Statement, General Tommy R. Franks, commanding U.S. Central Command, before the Senate Committee on the Armed Services, Washington, DC, September 19, 2000. Prior to Operation Northern Watch, the United States enforced a no-fly zone over northern Iraq through Operation Provide Comfort.

tive weapons."[38] In an address to the United Nations General Assembly in September 2002, President Bush said that Iraq continued to support terrorists and that al Qaeda members had escaped from Afghanistan to Iraq. He demanded that Iraq destroy its WMD, end support for terrorism, and cease persecution of its civilian population.[39] On March 17, 2003, President Bush said:

> It [the Iraqi regime] has a deep hatred of America and our friends. And it has aided, trained and harbored terrorists, including agents of al Qaeda. The danger is clear: using chemical, biological or, one day, nuclear weapons, obtained with the help of Iraq, the terrorists could fulfill their stated ambitions and kill thousands of innocent people in our country, or any other.[40]

In the same speech, President Bush presented an ultimatum: Saddam Hussein and his sons must leave Iraq within 48 hours or the United States would initiate military operations. Hussein defied this ultimatum, saying that war against the United States would be "the decisive battle between the army of faith, right, and justice, and the forces of tyranny and American-Zionist savagery on the other [side]."[41] On March 19, President Bush announced the start of military operations "to disarm Iraq, to free its people and to defend the world from grave danger." He added that the United States would not "live at the mercy of an outlaw regime that threatens the peace with weapons of mass murder."[42]

---

[38] The President's State of the Union Address, The United States Capitol, Washington, DC, January 29, 2002, available at http://www.whitehouse.gov/news/releases/2002.

[39] President's Remarks at the United Nations General Assembly, New York, September 12, 2002, available at http:/www.whitehouse.gov/news/releases/2002.

[40] Remarks by the President in Address to the Nation, The Cross Hall, The White House, Washington, DC, March 17, 2003, available at http:/www.whitehouse.gov/news/releases/2002.

[41] Burns, 2003.

[42] Remarks by the President in Address to the Nation, The Oval Office, The White House, Washington, DC, March 19, 2003, available at http:/www.whitehouse.gov/news/releases/2002.

Operation Iraqi Freedom quickly toppled Saddam Hussein's regime at low cost to coalition forces. The operation began on March 19 with an unsuccessful attempt to kill Saddam Hussein using Tomahawk missiles and GPS-guided bombs dropped by F-117 fighters.[43] In sharp contrast to Operation Desert Storm, conventional ground operations began only one day later. The Army's 3rd Infantry Division (Mechanized), operating west of the Euphrates River, advanced more than 500 kilometers in three days against generally light resistance. During the first week of April, the 3rd Infantry Division forced the Karbala Gap and entered Baghdad. Simultaneously, the 1st Marine Division advanced east of the Euphrates River and subsequently continued to Tikrit, while Task Force Tarawa reduced resistance in An Nasiriyah. President Bush declared an end to hostilities on May 1. Despite some inefficiencies, Operation Iraqi Freedom was an outstanding example of joint operations, particularly the advantages to be derived from coordinating ground operations with air attacks against enemy ground forces. The Iraqi Republican Guards, which the United States expected would present the most serious challenge, offered only scattered resistance to coalition ground forces. Indeed, most resistance came from paramilitary units, which continued to be a problem long after the fall of Baghdad.

Operations

U.S. Central Command originally developed a plan to invade Iraq that would have required more than 500,000 troops, about the number used during Operation Desert Storm, but Secretary of Defense Donald Rumsfeld and his deputies insisted on a much smaller force. After several iterations, Gen. Franks finally produced a plan that required only 151,000 troops.[44] The plan envisioned invading Iraq from the south through Kuwait and from the north through Turkey. However, on March 1, 2003, during a session of the Turkish Parliament, members of the newly ascendant Justice and Development

---

[43] Sanger and Burns, 2003, p. 1; Risen, 2003.

[44] Galloway, 2003; Hersh, 2003.

Party surprised their own leaders by voting against an agreement that would have allowed large U.S. forces to base in Turkey and open a northern front in Iraq.[45] At this time, the Army's only digitized division, the 4th Infantry Division (Mechanized), was posed to disembark in Turkey. After being rerouted, the division began offloading in Kuwait a month later, several days after Baghdad had fallen. Turkey's refusal also affected plans to base coalition aircraft in Turkey and the flight routes for aircraft on carriers in the eastern Mediterranean Sea.

As a consequence, U.S. Central Command attacked with only one Army heavy division, the 3rd Infantry Division (Mechanized), in the main effort west of the Euphrates, while the 1st Marine Division conducted a supporting attack east of the river. The 101st Airborne Division (Air Assault) and a brigade of the 82nd Airborne Division followed behind the 3rd Infantry Division to exploit its success and to secure lines of communication. Thus, just two U.S. divisions attacked much larger Iraqi forces, assessed as totaling about 300,000 men. Regular Iraqi Army divisions melted away, much as they had done during Operation Desert Storm, but U.S. planners expected Republican Guards divisions defending Baghdad to offer resistance. While land forces paused for resupply in late March, air forces attacked the Republican Guards relentlessly. When the 3rd Infantry Division and the 1st Marine Division resumed the offensive on April 2, they found little left of the Republican Guards other than destroyed and abandoned equipment. They quickly overwhelmed Iraqi paramilitary forces equipped with light infantry weapons.

While conventional forces advanced along the Euphrates and Tigris rivers, SOF operated in western and northern Iraq. In western Iraq, U.S., British, and Australian SOF were inserted to hunt for ballistic missiles, which might again have been directed against Israel. They encountered so little resistance that they were able to seize and hold airfields for their own use. In northern Iraq, Army Special Forces and Air Force combat controllers reprised the role they had recently played in Afghanistan. The combat controllers called air

---

[45] Filkins, 2003; Graham, 2003.

strikes against Ansar al-Islam, a terrorist organization with ties to Iran, and against Iraqi regular forces deployed along the line of confrontation with Kurdish forces.

On March 26, some 1,000 paratroopers from the Army's 173rd Airborne Brigade parachuted from C-17 aircraft near an airfield in territory controlled by the Kurdistan Democratic Party. This force lacked enough combat power to launch an independent offensive, but it did buttress Kurdish defenses and perhaps discouraged Turkey from intervening.

During Operation Iraqi Freedom, Iraqi forces presented very little air defense apart from low-level ground fire, while attacks on enemy ground forces dominated the air effort. About two-thirds of the sorties were apportioned to counterland operations or unconventional warfare, which concentrated heavily on enemy ground forces. About 57 percent of the designated mean points of impact nominated during the operation were counterland targets. Almost 80 percent of those struck during the operation were counterland targets (carried as kill-box interdiction and CAS). Fixed counterland targets accounted for only 1 percent of those struck (see Table 3.2).

Because of basing constraints, longer-range bombers had predominated in Afghanistan. By contrast, shorter-range fighters flew most attack sorties in Operation Iraqi Freedom. The F/A-18 Hornet was by far the most numerous fighter (250 aircraft), followed by the F-16CJ (71), AV-8 Harrier (70), A/OA-10 (60), F-16 (60), F-14 (56), and F-15E (48). However, A-10 and F-15E aircraft were

Table 3.2
Counterland Targets During Operation Iraqi Freedom

DMPI Nominations (CL)		DMPI Struck (Fixed CL and KI/CAS)	
Number	Percentage	Number	Percentage
17,521	57.3	15,826	79.5

SOURCE: Moseley, 2003.
NOTE: CL = counterland; DMPI = designated mean point of impact; KI/CAS = kill-box interdiction/CAS.

engaged disproportionately to their numbers. Bombers included the B-52 (28 aircraft), B-1 (11), and B-2 (4). Because of Turkey's refusal to allow overflight, aircraft on carriers in the Mediterranean initially required aerial refueling by Air Force tankers. More than two-thirds of all expended munitions were laser-guided bombs, with the GBU-12 predominating. In addition, A-10 aircraft expended over 300,000 rounds of 30-mm ammunition, often during close combat when enemy forces had to be engaged very close to friendly forces. Operation Iraqi Freedom was the combat debut of sensor-fuzed weapons, cluster munitions that use infrared sensing and explosively formed penetrators to destroy concentrations of armored vehicles (see Table 3.3).

During the invasion of Iraq, some control measures worked well, while others were problematic. "Stacks" (airspace set aside for attack aircraft awaiting missions) helped make CAS timely and unrelenting. The kill-box system helped concentrate aircraft where they were most needed, especially when performing interdiction. There were few fratricide incidents, and the worst (at An Nasiriyah on March 23) occurred because land forces lost track of their own locations. However, coordination between the Combined Force Land Component Commander (CFLCC) and V Corps, on the one side, and the Combined Force Air Component Commander (CFACC) and the Combined Air Operations Center, on the other side, suffered from a lack of mutual understanding that hampered air operations,

Table 3.3
Selected Munitions Expended During Operation Iraqi Freedom

TLAM		Maverick		WCMD		LGB		JDAM	
%	Number	%	Number	%	Number	%	Number	%	Number
4	802	4.6	918	4.5	908	43.7	8,716	32.8	6,542

SOURCE: Moseley, 2003.
NOTE: TLAM = Tomahawk land attack missile; WCMD = wind-corrected munitions dispenser (includes CBU-103/105 SFW/107); LGB = laser-guided bomb (includes GBU-12/16/24/27/28 and EGBU-27); JDAM = joint direct-attack munition (includes GBU-31/32/35/37). The selected munitions account for 89.6 percent of those expended. Other munitions included the Hellfire (AGM-114), the high-speed anti-radiation missile (AGM-88), and guided weapons released by U.K. aircraft.

especially at the outset of the campaign. The CFLCC initially placed the fire support coordination line (FSCL) 140 kilometers beyond friendly forces. Short of this line, V Corps considered all kill boxes closed unless opened by the ASOC supporting the corps. Unfortunately, this center lacked the resources and equipment, especially communications equipment, to manage such a large area efficiently. The 1st Marine Expeditionary Force and the 1st Marine Division took a much different approach. The Marines defined a battlefield coordination line much closer to friendly forces and opened all kill boxes beyond this line, an approach that promoted more efficient use of air power.[46]

On May 1, three weeks after the fall of Baghdad, President Bush declared that "major combat operations in Iraq have ended."[47] However, opposition forces continued to mount attacks on the coalition, especially north of Baghdad in a region favored by Hussein's regime and largely populated by Sunni Moslems. In June, a Shi'ite mob massacred six British soldiers who were training Iraqi policemen.[48] These early attacks were small-scale and were usually conducted with explosive devices and light infantry weapons, including rocket-propelled grenades. On July 22, 2003, acting on information received from an Iraqi, elements of the 101st Airborne Division surrounded a house where Hussein's sons Uday and Qusay were hiding. After a siege lasting several hours, U.S. forces assaulted the house and killed the brothers in a firefight. During late summer 2003, the targets of Iraqi insurgent attacks widened to include Iraqis who cooperated with U.S. authorities, critical infrastructure, and the UN mission. On August 20, a truck bomb exploded near the UN headquarters in Baghdad, killing the UN special envoy, Sergio Vieira de Mello. The U.S. military also had to contend with criminals bent on looting and extor-

---

[46] This discussion is based on unpublished RAND research conducted by Forrest Morgan on Operation Iraqi Freedom.

[47] Sanger, 2003.

[48] Banerjee, 2003; Chandrasekaran, 2003; McGrory and Evans, 2003.

tion.[49] On December 18, 2003, Task Force 21 and elements of the 1st Brigade, 4th Infantry Division (Mechanized) captured Saddam Hussein near Tikrit in northern Iraq. The ex-dictator was found hiding in a "spider hole" barely large enough to lie in at full length, and he appeared disoriented.

Following the capture of Hussein, the level of violence declined for a time, but it increased greatly during spring 2004. In April and May, U.S. forces conducted large-scale operations against Sunni insurgents in Fallujah and Shi'ite militia in Najaf. The previous year, the 82nd Airborne Division had ceased patrolling Fallujah in order to avoid provoking resistance. When the 1st Marine Division assumed responsibility for the area, it began preparing a deliberate operation to suppress the insurgency in Fallujah while winning the support of its inhabitants. But on March 31, 2004, gunmen killed four employees of Blackwater Security Consulting who were providing security to a convoy. Their vehicles were set on fire, and a mob subsequently dragged charred bodies through the street and hung them from a girder bridge. In response, the United States sent Marines into Fallujah, but after a few days of fighting, they halted the offensive and negotiated with the insurgents. The negotiations produced a local force, termed the Fallujah Brigade, which failed to restore order in the city. At the end of his tour, the commander of the 1st Marine Expeditionary Force, Lt. Gen. James T. Conway, said that a more deliberate operation might have succeeded and that the sudden attack made the United States appear to be seeking revenge for the Blackwater incident. But once the United States went on the offensive, it should not have backed down. "Once you commit, you got to stay committed."[50]

During the invasion of Iraq in April 2003, the 101st Airborne Division entered the pilgrimage city of Najaf forcefully, while treating

[49] Fattah, 2003; Dixon, 2003.

[50] Chandrasekaran, 2004.

the shrine of Ali Abi Tlib[51] with great circumspection. Subsequently, Shi'ites resumed making pilgrimages to the shrine in large numbers, causing a return of prosperity. Incidents also occurred in Najaf, but there was not the widespread, continual resistance that characterized Fallujah until spring 2004, when radical Shi'ite cleric Muqtada al-Sadr ordered the Mahdi Army, a Shi'ite militia equipped with light infantry weapons, to secure the shrine of the Imam Ali. A Marine Expeditionary Unit and two Army battalions fought the Mahdi Army in the northern cemetery and the old city for several weeks. Ostensibly, U.S. forces were acting at the request of the Iraqi interim government, which conducted fruitless negotiations with Muqtada al-Sadr. The interim government was preparing to storm the shrine when the Grand Ayatollah Ali al-Sistani reached an agreement with Muqtada al-Sadr on August 27, 2004.[52] Al-Sadr's followers, carrying their weapons, withdrew from the shrine and returned to other areas, including the sprawling slum in northern Baghdad called Sadr City after Muqtada's father.

During the fighting in Fallujah and Najaf, air power provided surveillance, reconnaissance, and precision strikes. Predator was widely employed to provide near-real-time video of areas of interest. Army and Marine attack helicopters were continually used despite their vulnerability to gunfire, particularly in built-up areas, and AC-130 gunships were employed at night over areas of intense fighting. Fighters provided CAS and struck a variety of fixed targets, including the suspected hiding places of opposition leaders.

Insights

Air attacks were a major killer in Iraq, not only in open terrain, but in urban terrain as well. Indeed, air attack was often the preferred mode of attacking urban targets because it was potentially more precise and therefore presented less risk of collateral damage. When the 3rd In-

---

[51] The Imam Ali Abi Tlib, the most revered figure in the Shi'ite religion, was murdered in neighboring Kufa in 661. His shrine is contained in a magnificent gold-domed mosque in the center of Najaf. A vast Shi'ite cemetery borders the shrine on the north.

[52] Filkins and Burns, 2004.

fantry Division and the 1st Marine Expeditionary Force began their assault on Baghdad in the first week of April, they encountered almost no resistance from Republican Guards divisions, which had been deployed to protect the Iraqi capital. Personnel assigned to these divisions deserted *en masse*, apparently because of losses suffered from air attacks and probably also because they believed that they could not prevail in battle. Lightly armed paramilitary and militia repeatedly fought U.S. forces, but their attempts often bordered on being suicidal. In a typical engagement, Iraqi paramilitary forces would open fire suddenly with machine guns and rocket-propelled grenade launchers but would fail to inflict much damage. U.S. ground forces would quickly return much heavier fire with direct-fire weapons, fixing the Iraqis for subsequent destruction by mortars, artillery, and air attacks. Early in the campaign, Iraqi artillerists frequently opened fire on U.S. forces; U.S. artillery usually had the counterbattery mission because attack aircraft could not respond quickly enough.

The CAS process was effective in Iraq but still fell short of its potential. It was often difficult to get an aircraft with appropriate munitions on the target quickly enough to meet the needs of the ground forces. Weapon loads sometimes turned out to be inappropriate—for example, precision weapons were sometimes provided when the target was fielded enemy troops. Using the Army OH-58D Kiowa to lase targets for other aircraft could be frustrating unless both pilots had practiced this technique. Very often, TACs had to talk aircraft onto their targets in traditional fashion, which caused delay. The skills of controllers and pilots affected times required for a talk-on, but five to ten minutes was usually required to attain the necessary confidence that the pilot had identified the target correctly and that his attack would not endanger friendly forces. This process was complicated not only by differences in perspective, but also by differences in the mode of sensing. The target-verification time was in addition to whatever time was required to validate the request for air support and to bring the aircraft into contact with forward controllers. As a result, aircraft usually responded more slowly than mortars and artillery did, sometimes slowly enough to try the patience of battalion and brigade commanders. Land forces especially esteemed AC-130 and

A-10 aircraft for their ability to loiter in the target area and their crews' familiarity with CAS tasks and procedures.[53]

TACs were critical to the success of joint operations in Iraq, as they had been previously in Afghanistan. Sharing the risks of infantrymen, TACs understood tactical situations on the ground and how air power could be best applied. They displayed not only the technical expertise required to call strikes close to friendly troops, but also the judgment to determine when attacks should be withheld to avoid fratricide. During engagements in built-up terrain, TACs often began by calling attacks at targets 400 to 600 meters beyond friendly troops and then brought them closer, much as an artillery observer would "walk" rounds. During the Vietnam War, controllers habitually flew in Army helicopters, but the technique fell into disuse as the lessons of that war were forgotten. In Iraq, the 101st Airborne Division put TACs in the right-hand seat of OH-58D Kiowa helicopters during some of its armed reconnaissance missions. Flying in the Kiowa allowed controllers to move about quickly and to see the battlefield from the perspective of an attack pilot. This perspective is important because landmarks appear more or less prominent depending upon the altitude and speed of the observer.

Operation Iraqi Freedom demonstrated the vulnerability of attack helicopters to ground fire during deep operations. The Army and the Marine Corps initially had very different tactics for attack helicopters. Broadly speaking, the Army emphasized deep attack with its AH-64D Apaches, while the Marine Corps emphasized close support with its less-capable AH-1W Cobras. Just before dawn on March 24, the Army's 11th Aviation Regiment conducted a large-scale attack on elements of the Iraqi Medina Division near Karbala. As the Apaches approached their targets, they came under intense ground fire from every direction and were compelled to curtail their mission. Thirty Apaches returned with combat damage, but only one went down in

---

[53] Interview with Captain Jon E. Chasser and Captain Marco Parzycn, Air Liaison Officers, 15th Air Support Operations Squadron, Fort Stewart, Georgia, October 28, 2003.

enemy-held territory.[54] After this experience, Army forces employed Apaches more cautiously to reduce their exposure to ground fire. The Army is currently reviewing its doctrine for employment of attack helicopters.[55] It may see greater need for joint air attack conducted in partnership with fixed-wing attack aircraft.

UAVs expanded their role during operations in Iraq. They were especially useful in providing targeting information for air strikes. The RQ-4 Global Hawk, which had played only a marginal role in Afghanistan, transmitted continuous near-real-time images to the Combined Air Operations Center at Prince Sultan Airbase in Saudi Arabia and to V Corps headquarters.[56] The Army operated Hunter and the Marine Corps operated Pioneer unmanned vehicles. Both services became convinced that they require UAVs organic to tactical formations. As these vehicles become available, they will provide invaluable data for artillery fire support and air attack and will also initiate attacks with their own on-board weapons. Eventually, UAVs may begin to supplant manned helicopters in low-level armed reconnaissance and to deliver supplies to ground forces.[57]

Operation Iraqi Freedom proved the continuing relevance of heavy armor, not only in open country, but also in urban warfare. M-1 Abrams tanks took first-round hits, usually from cannon and rocket-propelled grenades, without suffering serious damage. Their ability to survive hits was critically important because the enemy often went undetected until he opened fire. "Thunder runs" into cities quickly disrupted Iraqi defenses, preventing protracted urban combat that could have been costly. Based on this experience, the Army might conclude that it will need heavy armor for the foreseeable future. Although almost invulnerable to most infantry weapons, the Army's heavy forces will continue to require a close partnership with attack aircraft, especially during fluid, fast-moving operations and

---

[54] Sheridan, 2003; Gordon, 2003.

[55] Trimble, 2003. The reference is to Gen. John M. Keane, Vice Chief of Staff of the Army.

[56] Scarborough, 2003.

[57] Kaufman, 2003; Fulghum, 2003.

combat in built-up areas. To make this partnership work, Army forces will need access to reconnaissance data at very low echelons and will also need the ability to transmit precise targeting data.

Changes to Doctrine

All three of these recent operations suggest that counterland doctrine should be improved. In Kosovo, attacks on fielded forces were termed "close air support" even though there were no friendly forces to support, simply because the procedures resembled those for CAS. In Afghanistan, air forces attacked the Taliban in ways that stretched doctrine. Some attacks were obviously close support of friendly forces, sometimes to help these forces break through Taliban defenses and sometimes to keep them alive. Other attacks were more like interdiction, and some seemed to have no relationship to friendly forces at all.[58] In Iraq, air attacks were primarily intended to achieve effects that would facilitate land operations intended to topple the regime. Specifically, air attacks were expected to fix Iraqi forces in southeastern Iraq, reduce the combat power of Iraqi forces defending Baghdad, and help Kurdish forces advance through the "Green Line" in northern Iraq. In addition, air forces constantly flew CAS for U.S. land forces, especially those advancing through the Tigris-Euphrates corridor. In this case, doctrinal definitions seemed adequate and appropriate, but control measures caused difficulty.

Current joint doctrine is deficient in its definitions of those missions termed "counterland" by the Air Force and in the associated control measures, especially the FSCL. Missions should be redefined in terms of air power's relationship to land and naval power. Ultimately, the FSCL should be replaced with an area concept that is more suitable for the rapid, fluid operations envisioned in joint and

---

[58] To resolve this difficulty, Maj. Gen. Deptula suggested creating a new category called "battlefield air operations," defined as operations against enemy ground forces where friendly ground forces are not present or engaged in operations in direct support of air operations. See Deptula, 2003.

service doctrine, including the Army's transformation initiatives. These changes should promote efficiency and timeliness without sacrificing flexibility.

In addition, there should be as much commonality as possible in terminology, control measures, and procedures across the services, since they all operate combat aircraft, and all except the Army operate fixed-wing attack aircraft. Each service has developed its own way to control combat aircraft, including different control entities and different qualifications for personnel performing the terminal-attack function. The goal should be commonality to assure that pilots and controllers encounter no disjuncture or misunderstanding when working across all four services.

Missions

Current joint doctrine recognizes two counterland missions: CAS and AI. "Close air support is air action by fixed- and rotary-wing aircraft against hostile targets that are in close proximity to friendly forces and that require detailed integration of each air mission with the fire and movement of those forces."[59] "Interdiction is an action to divert, disrupt, delay, or destroy the enemy's surface military potential before it can be used effectively against friendly forces."[60] Doctrine recognizes that interdiction can occur within the area of operations of land and amphibious force commanders or outside their areas. If it occurs within their areas, these commanders are responsible for synchronization of organic fires and air interdiction. If it is outside their areas, the joint-force air-component commander conducts the effort under direction of the joint-force commander (see Table 3.4).

Joint doctrine also identifies the mission of "strategic air attack" but does not define it.[61] DoD has no definition for "strategic air attack," but it offers this definition for "strategic air warfare":

---

[59] Chairman, Joint Chiefs of Staff, 2003c, p. I-1.

[60] Chairman, Joint Chiefs of Staff, 1997a, p. v.

[61] Chairman, Joint Chiefs of Staff, 2003a, p. II-3. However, joint doctrine does define "joint strategic attack." (JP 3-0, p. GL-12).

Air combat and supporting operations designed to effect, through the systematic application of force to a selected series of vital targets, the progressive destruction and disintegration of the enemy's war-making capability to a point where the enemy no longer retains the ability to wage war. Vital targets may include

Table 3.4
Current Doctrine for Counterland

<table>
<tr><th>Type</th><th>Mission</th><th>Area</th><th>Control</th></tr>
<tr><td>CAS</td><td>"Air action by fixed- and rotary-wing aircraft against hostile targets that are in close proximity to friendly forces and that require detailed integration of each air mission with the fire and movement of those forces" [JP 1-02].</td><td>"CAS can be conducted at any place and time friendly forces are in close proximity to enemy forces. The word "close" does not imply a specific distance; rather, it is situational. The requirement for detailed integration because of proximity, fires, or movement is the determining factor" [JP 3.09.3, p. I-2].</td><td>"Commanders employ CAS to augment supporting fires to attack the enemy in a variety of weather conditions, day or night" [JP 3-09.3, p. I-3]. "Terminal attack control of CAS assets is the final step in the TACS for CAS execution. There are both ground and air elements of the TACS to accomplish this mission" [JP 3-03.9, p. II-9].</td></tr>
<tr><td>Int. AO</td><td rowspan="2">"Air operations conducted to destroy, neutralize, or delay the enemy's military potential before it can be brought to bear effectively against friendly forces at such distance from friendly forces that detailed integration of each air mission with the fire and movement of friendly forces is not required" [JP 1-02].</td><td>"Interdiction operations within AOs occur simultaneously with joint interdiction operations ranging theater- and/or JOA-wide." "Interdiction can occur both short of and beyond the FSCL" [JP 3-03, p. ix].</td><td>"As supported commanders within their area of operations (AO), the land and naval force commanders are responsible for synchronizing maneuver, fires, and interdiction" [JP 3-03, p. ix].</td></tr>
<tr><td>Int. JOA</td><td>"Theater- and/or joint operations area (JOA)-wide interdiction operations may be planned and executed by the JFC staff or the appropriate commander as directed by the JFC" [JP 3-03, p. viii].</td><td>The JFACC accomplishes the mission of air interdiction and may function as the supported or supporting commander as designated by the JFC [JP 3-30, p. II-2].</td></tr>
</table>

NOTE: AO = area of operations; CAS = close air support; JFACC = joint-force air-component commander; JFC = joint-force commander; JOA = joint-operations area; JP = joint publication; TACS = theater air-control system.

> key manufacturing systems, sources of raw material, critical material, stockpiles, power systems, transportation systems, communication facilities, concentration of uncommitted elements of enemy armed forces, key agricultural areas, and other such target systems.[62]

The Air Force has promulgated a doctrine for strategic air attack, which should eventually enter joint doctrine. The Air Force definition reads:

> Military action carried out against an enemy's center(s) of gravity or other vital target sets, including command elements, war-production assets, and key supporting infrastructure in order to effect a level of destruction and disintegration of the enemy's military capacity to the point where the enemy no longer retains the ability or will to wage war or carry out aggressive activity.[63]

This definition omits mention of enemy forces, but in the accompanying text they are offered as centers of gravity, with Iraq's Republican Guards cited as an example.[64] Examples of "strategic air attack" are drawn from World War II, the Vietnam War, and Operation Desert Storm.

In current joint doctrine, only CAS is satisfactorily defined. Interdiction is poorly defined, and strategic air attack is only mentioned.

The definition for interdiction is unsatisfactory, if only because it is far too broad. It embraces every attack against enemy forces beyond CAS, ranging from attacks in the area of operations to theater-level air attacks. In doing so, it obscures the crucial distinction between attacks that complement a plan of maneuver on the ground and attacks that are directed against an enemy's military potential independent of friendly maneuver. The latter is actually strategic air attack, a very different kind of effort that should not be confused

---

[62] Department of Defense, 2003.

[63] Air Force Doctrine Center, 1998, p. 52.

[64] Ibid., p. 18.

with interdiction. The FSCL would clarify matters if it delineated areas where different missions occur, but instead it bisects the area where interdiction occurs.[65] It is at best unhelpful and at worst confusing to draw a line so that interdiction occurs on both sides. Finally, the term "interdiction" is outmoded and misleading. It implies merely prohibiting an enemy from acting,[66] but air power can destroy an enemy's forces.

There is no definition of strategic air attack in joint doctrine. "Strategic air warfare" as defined by DoD resembles "strategic air attack" as defined by the Air Force, but there are several differences. The DoD definition targets only the enemy's ability to wage war, but the Air Force definition targets his ability and will, implying a psychological dimension. The lists of appropriate target sets also differ considerably. For example, the DoD definition omits command elements, while the Air Force definition omits enemy forces. In current doctrine, interdiction outside the areas of operations delineated for land- and amphibious-force commanders sounds vaguely like strategic air attack, but it would logically include only the counterland targets. TACs can, of course, support both AI and strategic attack.

Joint and service doctrine should reflect three basic missions for air power delineated by air power's relationship to land power and to naval power in joint operations. The first two fall almost entirely under counterland, while the third embraces a wide variety of targets. Expressed in simple language, the three missions should be:

1. Help defeat enemy land forces in close combat.
2. Weaken enemy land forces before close combat occurs.
3. Destroy an enemy's ability to conduct warfare.

---

[65] The land- or amphibious-force commander establishes the FSCL. Forces attacking beyond the FSCL must inform affected commanders in time to avoid fratricide. The land- or amphibious-force commander controls all air-to-ground and surface-to-surface attacks short of the FSCL, which should follow well-defined terrain features. See Chairman, Joint Chiefs of Staff, 2001c, III-42–III-44, GL 9–10.

[66] "Interdiction" literally means to forbid or prohibit, from the Latin *interdicere* (speak between). The idea was that air attacks would prohibit enemy forces from closing with friendly forces.

The first mission would approximate today's CAS mission but is more limited. This mission should be confined to an area of close combat where detailed integration is imperative. This area should extend from the forward line of own troops to the depth of close combat on land. Normally, that depth would be the range of the land force's organic indirect-fire weapons, typically about 30 to 40 kilometers. But in some situations, e.g., when an air-mobile assault, amphibious assault, or special operation was under way, it could have much greater depth. Within the area of close combat, attacks by fixed-wing and rotary-wing aircraft should be integrated with direct fire by automatic weapons, cannons, and missiles and with indirect fire by mortars, artillery, and rockets. Increasingly, air attacks should be integrated with reconnaissance and strike missions flown by UAVs. Maneuver (land and amphibious) force commanders should have authority to request air attacks. They should integrate actions of their own assets with air attacks, normally on the basis of advice received from air liaison officers (ALOs) and TACs. Air attacks should normally require terminal-attack control, exercised by ground-based or airborne FACs, to achieve the required precision and timeliness. Crews of airborne control platforms, specially trained fighter pilots, and TACs aloft in helicopters might conduct forward air control. The term "close air support" is acceptable, but a more appropriate term would be "close air attack." The word "support" is bland and even misleading if taken to imply a passive role for air power.

The second mission would cover today's mission of interdiction, less strategic air attack, while being more closely defined. This mission should be performed in an area that extends from the area of close combat to the depth of planned maneuver by friendly ground forces.[67] In this area, a joint air-forces commander should be responsible for conducting an air effort that degrades enemy forces before

[67] During the Cold War, the U.S. Army Training and Doctrine Command with concurrence from the Air Force Tactical Air Command promulgated doctrine for "battlefield air interdiction" (BAI) that closely resembled the proposed mission. BAI required joint coordination during planning because it had near-term effects on friendly forces, but it was controlled and executed by an air commander. For a discussion of BAI and its demise, see McCaffrey, 2002.

they come within range of friendly forces. He could effect this degradation by attacking enemy lines of communication and stocks of war material, as well as by destroying enemy forces. He should have authority to conduct attacks without prior coordination and without terminal control. These attacks should complement the joint plan of maneuver and focus on achieving the joint-force commander's desired effects, but how closely they are constrained by this plan should vary according to the situation. For a protracted "softening-up" effort in great depth, the constraint might be loose, e.g., spare the bridges because we will need them. For an effort conducted days or even hours before the start of maneuver, the constraint might be tight, e.g., concentrate on destroying this enemy corps because it presents the greatest threat. But whether loosely or tightly constrained, the air effort should complement the plan of maneuver and not be an unrelated effort. Depending on the situation, an air-force commander might exert control through a theater-level air operations center (AOC) or through a subordinate AOC, collocated with a maneuver force. The term "interdiction" might be replaced by a more descriptive term, such as "deep attack."[68]

The third mission is strategic air attack, an effort to reduce the enemy's war-making potential unconstrained by a plan of ground maneuver. It might take place in areas beyond the depth of planned maneuver or in theaters where no maneuver is contemplated. Possible target sets should include enemy forces among leadership, communications, power-generation, transportation, dual-use industry, and infrastructure targets. Most of the strategic bombing conducted against Germany and Japan in World War II, most of the bombing done in North Vietnam, and most of the air effort against Serbia outside of its Kosovo province fall into this mission. It should long since have been

---

[68] Current doctrine for Army attack helicopters makes the useful distinction between "close operations" against enemy forces in contact with friendly ground forces and "deep operations" against enemy forces not in contact. See, for example, Headquarters, Department of the Army, 1997, paragraphs 3-19 and 3-20.

recognized in joint doctrine. An air-component commander should normally direct strategic air attack under guidance from a joint-force commander. Table 3.5 summarizes these three missions.

Fire Support Coordination Line

Under current doctrine, the FSCL is unrelated to missions; indeed it lies across the area where interdiction is performed. The FSCL replaced the bomb line used to avoid fratricide during World War II. The bomb line was usually placed at about artillery range, but under heavy pressure, a land-force commander might bring it within a few hundred yards of friendly positions, accepting greater risk of fratricide to gain more-effective air attacks. The bomb line was thus more clearly defined and closely held than the FSCL, which has been cast adrift of any clear rationale.

Table 3.5
Proposed Doctrine for Counterland

Mission	Area	Control
Help defeat enemy land forces in close combat.	Close combat area, extending from forward line of own troops to depth of friendly land-force actions.	Maneuver force commanders integrate organic means and air attacks. Air attacks are at their request. Terminal control by a ground or airborne FAC is required.
Weaken enemy land forces before close combat occurs.	Maneuver area, extending from the close-combat area to the limit of planned maneuver.	Air-force commander directs air attacks to complement the plan of maneuver. Prior coordination and terminal control are not required. Control through an AOC.[a]
Destroy an enemy's ability to conduct warfare.	Theater of military operations less the maneuver area, if any.	Air-component commander directs air attacks under guidance of joint-force commander. Control through theater JAOC.[a]

NOTE: FAC = forward air controller; AOC = air operations center; JAOC = joint air operations center.

[a] TACs might also support these missions.

Placement of the FSCL is often contentious because doctrine says only that it should "strike a balance,"[69] and the services disagree about where this balance should lie. During Operation Desert Storm and Operation Iraqi Freedom, Army commanders usually established the FSCL in operational depth, i.e., to the depth of a corps commander's interest. From a land perspective, this placement made sense because it assured that air attacks and the plan of maneuver would be complementary. But from an air perspective, it was ill advised. It required maneuver force commanders to control air attacks where there was no risk of fratricide and no need to integrate fires. Such control was at best superfluous and could have imposed costly delays.

According to current doctrine, the FSCL "should follow well-defined terrain features." This doctrine reflects an older state of affairs, before inertial navigation and the GPS were introduced, when military forces were uncertain of their positions and easily became lost. Because of this uncertainty, control measures had to be firmly anchored on recognizable terrain features. But today, every aircraft and nearly all soldiers have access to accurate geopositioning. Moreover, today's air operations are typically controlled through artificial reference systems such as kill boxes based on latitude and longitude. The status of kill boxes and their subdivisions tells airmen what actions are permissible within the areas they define. In many applications, kill boxes are a more efficient way to delineate battle space than traditional lines, especially during fast-paced, fluid operations like those envisioned under current programs to transform military forces.[70]

---

[69] "Placement of the FSCL should strike a balance so as not to unduly inhibit operational tempo while maximizing the effectiveness of organic and joint force interdiction assets. Establishment of the FSCL too far forward of friendly troops can limit the responsiveness of air interdiction sorties and could unduly hinder expeditious attack of adversary forces" (Chairman, Joint Chiefs of Staff, 2001c, p. III-44).

[70] Kill boxes normally cover the entire area of operations. Commanders may also designate areas where particular permissions and restrictions apply, including free-fire areas, no-fire areas, and restrictive-fire areas.

The FSCL should be redefined so that it bounds the area where some mission is performed, but which mission? Soldiers would presumably tend to place the FSCL at the outer edge of interdiction, to assure that air attacks complement planned maneuver. Airmen would be more likely to place it at the outer edge of CAS, to assure that air attacks are conducted with timeliness and efficiency. Both concerns are valid, and they are not mutually exclusive. The solution is to place the FSCL at the outer edge of CAS but establish the principle that air attacks in any area where maneuver is planned will complement that plan. Land-force commanders should trust air-force commanders to observe that principle. If their opinions diverged too sharply, either could appeal to the joint-force commander, whose decision would be binding.

The next step would be to replace the FSCL with a common grid reference system akin to kill boxes that offers greater precision and flexibility. At the current time, fielded equipment may be too unreliable and incompatible to completely replace the FSCL and support a common grid reference system across the joint force. But very soon such a system should become feasible (see Figure 3.1).

On the digital battlefield envisioned for the future, Army and Air Force staff-level officers working together in the ASOC would open and close grids as needed. These actions would instantly and automatically be disseminated to the displays of all battle-management, intelligence, surveillance, and reconnaissance (ISR), and strike aircraft; to command centers; and to maneuver forces. Boxed, coded close combat would contain friendly ground forces and require that strike aircraft be controlled by TACs. All strikes in the boxes would be authorized by the ground commander. The grids in the maneuver area would contain no friendly ground forces and would allow air to operate without terminal control, but all strikes would be integrated with the planned ground scheme of maneuver. The rest of the theater of military operations would be coded free-fire or no-fire zones, as appropriate.

Figure 3.1
Replacing the FSCL with Kill Boxes

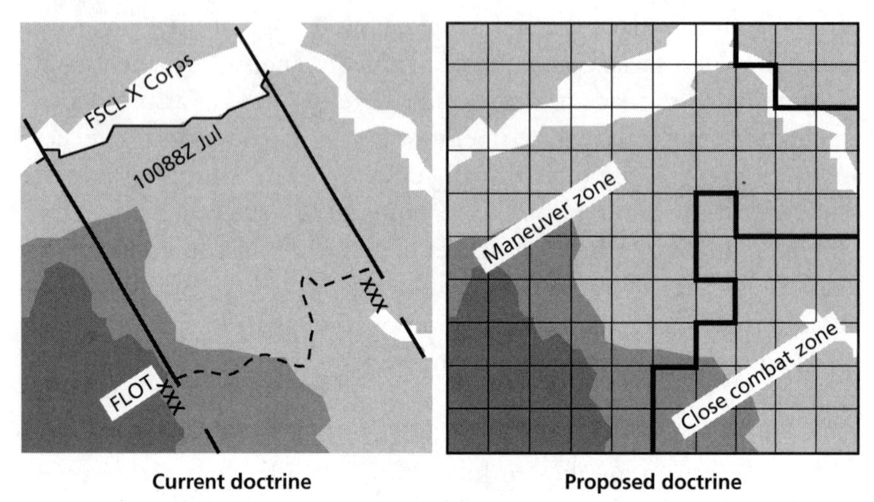

RAND *MG301-3.1*

Supported and Supporting Commanders

In joint doctrine, a superior commander at any level may establish support relationships among his subordinates. The President and his Secretary of Defense normally establish support relationships among the unified commands to plan and execute military operations and campaigns. A joint-force commander may establish support relationships among his subordinates, including his functional-component commanders, such as the land-component commander and the air-component commander. A supporting force is expected to "aid, protect, complement, or sustain" another force.[71] A supported commander has authority to "exercise general direction of the supporting effort."[72] A supporting commander "determines the forces, tactics, methods, procedures, and communications"[73] to be used in providing support. In the case of CAS:

---

[71] Chairman, Joint Chiefs of Staff, 2001b, p. III-9.

[72] Ibid.

[73] Ibid.

> Supporting activities can take many forms as air, land, sea, special operations, and space forces support one another. For instance, support occurs when the supporting force acts against targets or objectives that are sufficiently near the supported force to require detailed integration or coordination of the supporting attack with fire, movement, or other actions of the supported force.[74]

At first glance, it seems unobjectionable to describe air attacks in close proximity to friendly land forces as "support." Indeed, it is axiomatic that a ground-force commander must have the authority to control all fires and air attacks on his battlefield. But the word is ultimately inappropriate: It goes too far and not far enough. It goes too far in suggesting that the ground-force commander should have a scheme of maneuver that the air-force commander merely assists, as though ground maneuver should be planned without considering the effects of air power. In fact, a ground-force commander needs to consider how both kinds of combat power will combine to achieve his objective. Depending upon the situation, he may want one or the other to be dominant at any given time. Every tactician understands that fire and maneuver enable each other. To an increasing degree, especially for the Army's light forces, maneuver and air attack will enable each other, and they need to be thought out together.

At the same time, the word "support" doesn't go far enough in stating an ultimate relationship between air power and land power that becomes apparent *in extremis.* When survival becomes the issue, air forces will always assist ground forces, even at very high risk to themselves. This imperative applies even when very small forces are involved. Consider the example of small SOF on a reconnaissance mission. They may have been inserted simply to find targets for air forces to attack, and in this sense they support air forces in the deep attack mission. But if these special operators come under pressure, air forces will do everything possible to protect them with close attack

[74] Chairman, Joint Chiefs of Staff, 1997b. See also Chairman, Joint Chiefs of Staff, 2003b, p. 511.

and, if necessary, to extract them. The special operators will be a supported force until the emergency has passed.

The supported and supporting relationships have become so firmly ingrained in doctrine and practice that it would be unrealistic to demand their removal. Moreover, they are very useful *among* unified commands; it is *within* unified commands that they tend to be less helpful. The most fruitful relationship between air power and land power is not for one to support the other, but rather for both to act in partnership. At any given time in a battle or campaign, air power or land power might predominate. The joint-force commander oversees this partnership and determines which partner should play the predominant role at any given time. Geographic lines drawn across the battle space should not be allowed to define these roles. From a command-and-control perspective, it is important to give the most appropriate commander the requisite authority to accomplish his assigned tasks. In the close-combat area, this will be a land-force commander, who must integrate air attacks by fixed- and rotary-wing aircraft with organic fires and maneuver. In areas where land forces are not present but maneuver is planned, this authority should be given to an air-forces commander. However, his effort should complement the plan of maneuver by creating conditions for its success. Finally, where no maneuver is contemplated, an air-forces commander should have freedom of action within a joint-force commander's guidance.

CHAPTER FOUR

Army Transformation and the Air-Land Partnership

Introduction

Over the course of the next decade, the U.S. Army plans to reform its tactical organizations, training, and equipment via a process it calls "Army Transformation."[1] The Army's reforms are driven by its perception of trends in the nature of ground warfare and, more proximately, by planning guidance from DoD.[2] The Army's aim is to fundamentally alter the way its units operate, and consequently the way they interact as part of the joint team.[3]

This chapter assesses the potential effect of Army Transformation on the air-ground partnership. It finds that Army Transformation is likely to significantly influence the employment of joint fires on future battlefields, particularly by reducing the amount of artillery fire support available to ground forces, with consequent implications for Air Force counterland operations. These trends appear to be enduring aspects of Army Transformation, transcending recent changes initiated by new Army Chief of Staff Gen. Peter J. Schoomaker, and in all likelihood enduring into the future as Army

[1] Headquarters, Department of the Army, 1999, 2003f; Headquarters, Department of the Army, "General Peter J. Schoomaker 35th Chief of Staff of the Army, Arrival Message," Washington, DC, August 1, 2003.

[2] Office of the Secretary of Defense, 2003.

[3] See U.S. Army Training and Doctrine Command, 2003, p. 2. See also "Remarks by General Peter J. Schoomaker, Chief of Staff of the Army Before the United States House of Representatives Committee on Armed Services," 2nd Sess., 108th Cong., January 28, 2004.

Transformation continues to evolve.[4] They therefore deserve the attention of Air Force planners.

The Changing Battlefield

Army Transformation is, in many respects, a straightforward adjustment to a changing operational context. The art and science of warfare are in constant flux, driven by changes in politics, technology, and the evolutionary pressures of military competition.[5] Currently, the U.S. approach to land warfare is driven by three mutually reinforcing trends: the emergence of precision weapons, the growth and proliferation of information technology (IT), and the post–Cold War prominence of power projection in U.S. defense planning.[6]

The Precision Revolution

By radically increasing the lethality of individual munitions, the advent of precision weapons is altering the tactical and operational approaches of the U.S. military and its adversaries. Armed with significant stocks of precision munitions and steadily improving reconnaissance capabilities, U.S. forces can destroy almost any identifiable target in almost any environment.[7] All the U.S. armed services are embracing precision and reshaping operational concepts, doctrine, and capabilities to better exploit precision capabilities.[8] The Army is fully participating in this move toward precision—its tactical concepts for the future battlefield increasingly emphasize standoff preci-

---

[4] Research for this chapter was completed before the 2003 change in Army leadership; the chapter was selectively updated through August 2004.

[5] In the extensive literature, see particularly Howard, 1970; McNeil, 1982; van Creveld, 1991; Jones, 1987.

[6] For a useful overview, see Davis and Shapiro, 2003. For an earlier view, see Bellamy, 1990.

[7] See Lambeth, 2000; Metz and Millen, 2003.

[8] The "capabilities push" described here is matched to a degree by a "requirements pull" as national leaders demonstrate increasing sensitivity to collateral damage. See Headquarters, United States Air Force, 2003; Headquarters, Department of the Army, 1999; Headquarters, Department of the Navy, 2002; Chairman, Joint Chiefs of Staff, 2001a.

sion engagement, rather than the traditional emphasis on maneuver and shock.[9]

This shift toward precision is not merely conceptual. The Army is also working to field the hardware necessary to make precision engagement a reality. Current precision munitions under development include the 120-mm Precision Guided Mortar Munition, the 155-mm Excalibur extended-range guided projectile, and the Guided Multiple Launch Rocket System.[10] The Army is also investing in command-and-control systems and reconnaissance systems that will enable future Army units to reliably contribute to precision engagement by the joint team.[11] Through these investments, the Army is seeking to equip itself to engage adversary forces with a variety of joint and Army precision munitions from standoff ranges.[12]

The precision revolution has also influenced tactics employed by U.S. adversaries. U.S. firepower increasingly compels adversaries, including conventional military units, paramilitaries, and irregulars, to develop tactics to avoid destruction by precision munitions. The dominance of U.S. forces may compel adversaries to abandon operational-level mass and maneuver, decline major engagements with U.S. forces, and consequently forsake operational victory as traditionally defined. Instead, adversaries will more often turn to asymmetric approaches.[13] Recent conflicts in Somalia, Kosovo, Afghanistan, and Iraq illustrate the use of insurgent strategies to escape destruction by U.S. firepower. Enemy forces "go to ground," i.e., dis-

---

[9] See, for example, Headquarters, Department of the Army, 2001a.

[10] Headquarters, Department of the Army, 2002a.

[11] Headquarters, Department of the Army, 2003a, p. 48; U.S. Army Training and Doctrine Command, 2001a.

[12] These developments represent increased interest in precision weapons, rather than a new interest. The Army has fielded precision capabilities for more than 20 years, for example, the M712 Copperhead laser-guided cannon round (1982) and the AGM-114 Hellfire antitank guided missile (1985). Furthermore, many of its direct-fire systems, including the 120-mm cannon on the M1 Abrams series main battle tanks and the TOW series of antitank guided missiles, are incredibly accurate.

[13] On adversaries adapting to precision warfare in particular, see Scales, 2001. On asymmetric responses generally, see Applegate, 2001; McKenzie, 2000; Meigs, 2003.

perse, hide, and conceal their identity, while attempting to inflict casualties on U.S. forces through sudden, brief ambushes. On the tactical level, this approach is attractive because it minimizes exposure to U.S. firepower. On the strategic level, it involves prolonging the conflict by refusing decisive battle while inflicting casualties on the United States to erode public support.[14]

Information Technology

The information revolution of recent decades has reshaped every aspect of society, from culture to politics.[15] Microprocessors and networks continue to approximate Moore's Law[16] and proliferate throughout society with remarkable momentum.[17] In the defense sector, many have argued that the growth and proliferation of IT will spark a "revolution in military affairs," or at least significantly new approaches to command and control of military forces.[18] DoD has ambitious plans for U.S. forces to conduct rapid, highly coordinated parallel operations facilitated by seamless digital command-and-control systems.[19]

The Army has been working to integrate advanced IT into its combat units for more than 30 years.[20] It has initiated a variety of

---

[14] Record, 2000, 2002; Mueller, 2000.

[15] There is an extensive literature on the information revolution. See Alberts and Papp, 1997, 2000, 2001; Tellis et al., 2000.

[16] In 1965, Gordon Moore observed that over the period 1959–1965, the number of transistors on a chip doubled each year. He predicted that processing power would continue to rise exponentially, making home computers possible. Ten years later, Moore predicted that the number of transistors on a chip would continue to double every two years, a prediction that was generally realized.

[17] Nygren, 2002; Adams, 1998; Gompert, 1998; Arquilla and Ronfeldt, 1997; Wolfowitz, 2002.

[18] The literature on this point is quite extensive. Early entries include Echevarria and Shaw, 1992–1993, and Mazaar, 1993. More-recent entries include Alberts, 2002; Darilek et al., 2001.

[19] See Office of the Secretary of Defense, 2003; Wolfowitz, 2002. Both mention a number of categories of technology DoD wishes to exploit, but IT stands out as the primary category.

[20] Stanley, 1998, p. 12; Bowman et al., 1989, Executive Summary and Part II.

digital command-and-control programs, many of which have unfortunately been abortive efforts, while others have been fielded—but only to mixed reviews.[21] The most ambitious such effort was the Force XXI initiative during the 1990s, which attempted to integrate a number of developmental programs into a single coherent digital battle command system.[22] Force XXI was the Army's major modernization effort of the decade; it sought to "digitize" units from platoon to corps.[23] Force XXI experimentation concluded in 1998 with uncertain results, and digitization has been limited to gradual modernization of III Corps, the 4th Infantry Division (Mechanized), and the 1st Cavalry Division.[24]

With the start of Army Transformation in late 1999, the digital command-and-control systems developed under the auspices of Force XXI reemerged as the centerpiece for the Army's new vision. The Army's transformed combat organizations are equipped with the Army Tactical Command and Control System (ATCCS), which includes digital command-and-control systems for each battlefield functional area, e.g., maneuver, fire support, and logistics.[25] These systems are tied together into a "tactical internet" by high-bandwidth voice, video, and data-exchange networks.[26]

To the extent that it works as advertised, the ATCCS will enable Army ground forces to execute new operational concepts by providing a common relevant operational picture. In the past, ground forces usually deployed in compact, linear formations of contiguous units. In the future, ground forces will increasingly deploy in dispersed, nonlinear formations, using digitized command-and-control systems

---

[21] Stanley, 1998, pp. 12–27.

[22] U.S. Army Training and Doctrine Command, 1994.

[23] Hannah, 1997.

[24] Headquarters, Department of the Army, 2003a, p. 27.

[25] U.S. Army Training Initiatives Office, available at http://www.tio-armytransformation.net/aepublic/abcs/atccs_ps.htm.

[26] U.S. Army Training and Doctrine Command, 2001b, pp. 14–15. See also "Testimony of Steven W. Boutelle," 2004.

to enable units to provide mutual support. Army units participating in Operation Iraqi Freedom demonstrated a nascent form of this capability.[27]

Trends suggest that the ability to conduct high-tempo, dispersed, nonlinear operations will be increasingly important for battlefield success. If adversaries disperse to avoid precision munitions, U.S. land forces will have to operate in a dispersed way to deny them sanctuary and force them to accept battle. The ability to coordinate dispersed operations rapidly and effectively through advanced IT is therefore likely to be a central aspect of future ground operations.

Force Projection

Strategic factors have also shaped Army planning over the past decade. The post–Cold War strategic environment erased the Army's long-standing assumption that major combat would most likely occur in regions where the Army maintained a large, long-term presence.[28] The contemporary strategic assumption, beginning with the fall of the Soviet Union, is that the Army might be called upon to conduct operations in far-flung theaters where it lacks such presence. Since 1993, operations in Somalia, Haiti, Bosnia, Kosovo, Afghanistan, and Southwest Asia have all been examples of force projection.[29]

For Army planners, force projection creates a requirement to make Army units more strategically deployable, given a lack of forward basing. Correspondingly, one of the primary thrusts of Army Transformation has been to improve the Army's ability to deploy over strategic distances.[30] The Stryker brigades are specifically designed to deploy rapidly over such distances.[31] The Army's new

---

[27] See Congressional Budget Office, 2003, Chap. 1.

[28] Romjue, 1997a, pp. 2, 119–120.

[29] Difficulties in deployment during the 1999 conflict in Kosovo may have spurred Army Transformation. See Nardulli et al., 2002.

[30] Headquarters, Department of the Army, 1999. For a more recent restatement, see U.S. Army Training and Doctrine Command, 2003, p. 2.

[31] Vick et al., 2003, Chap. 1; Peltz et al., 2003; U.S. Army Training and Doctrine Command, 2001a, p. 6.

equipment programs, particularly the Stryker armored vehicle and future combat system, are constrained by a requirement to be deployable on C-130 aircraft, which may prove unattainable.[32]

The Army's Vision of Transformation

Army Transformation is the latest in a long series of major reform efforts.[33] The Army has embraced several such efforts in the modern era, ranging from its post-Vietnam reorientation on the defense of Western Europe[34] to the Army of Excellence reforms that shaped the combat units the Army took to Operation Desert Storm[35] and the Force XXI initiative that occupied most of the 1990s.[36] Army Transformation is intended to move the Army from the current force created by these successive efforts to a future force capable of dominating any adversary in the new operational context shaped by precision weapons, IT, and strategic force projection. It is a comprehensive reform effort, embracing all aspects of the Army, including equipment, organization, and doctrine.

New Equipment

The Army is seeking to acquire new platforms and systems under the auspices of Army Transformation. The Stryker family of armored vehicles, equipping the new Stryker brigades, is a prominent example. The Army is currently developing and procuring ten variants of the Stryker, including the infantry combat vehicle (50-cal. machine gun or 40-mm grenade launcher), mobile gun system (105-mm cannon), mortar carrier (120-mm mortar), antitank guided missile carrier (tube-launched, optically tracked, wire-guided missile system), and

---

[32] Gordon and Orletsky, 2003, pp. 192–193.

[33] Hawkins and Carafano, 1997.

[34] Romjue, 1984.

[35] Romjue, 1997b.

[36] Romjue, 1997a.

variants for reconnaissance, command, fire support, engineering, ambulance, and nuclear-biological-chemical reconnaissance.[37] A Stryker weighs about 20 tons and can be carried in a C-130 for short distances; C-5 and C-17 aircraft can lift several Strykers over strategic distances. The Stryker is lightly armored—sufficient only to protect against heavy machine-gun fire—but with appliqué armor, it will withstand early-generation rocket-propelled grenades.

The Army is equipping Stryker brigades with several advanced systems in addition to the Stryker family of vehicles. The Stryker brigade's cavalry squadron has tactical UAVs for reconnaissance and surveillance duties and Q-36 and Q-37 target-acquisition radars for counterbattery duties.[38] The Stryker brigade's sensor platoons are equipped with the remote battlefield sensor system, an array of passive seismic, acoustic, thermal, and magnetic unattended sensors. The sensor platoons also operate ground search radars and tactical signals intelligence systems.

Perhaps the most important technologies in the Stryker brigade are its suite of digital command-and-control systems. Each system brings new digitized capabilities to the brigade:

- The Force XXI Battle Command Brigade and Below (FBCB2) System disseminates a common operational picture to each vehicle and headquarters in the brigade. It also forms the backbone of the Blue Force Tracker System for combat identification and digital communications between vehicles and units.
- The Maneuver Control System allows brigade elements to plan collaboratively while dispersed throughout the combat zone. It is the brigade staff's primary means of developing and disseminating digital operations orders.

---

[37] "Stryker Family of Vehicles," Fact Sheet, Washington, DC: General Motors General Dynamics Land Systems Defense Group, November 2001, available at http://www.army.mil/features/stryker/stryker_spec.pdf.

[38] U.S. Army Training and Doctrine Command, 2001b, pp. 46–47.

- The All-Source Analysis System coordinates intelligence collection by brigade elements and allows the brigade staff to pull data from national and theater systems. Primarily the province of the brigade S-2 section, this system is also distributed to other brigade elements, notably the cavalry squadron and its surveillance systems.
- The Advanced Field Artillery Tactical Data System allows the brigade fires-effects coordination cell to coordinate fire support digitally throughout the brigade area. It automatically tracks, sorts, and stores fire support coordination mechanisms, fire plans, calls for fire support, and other critical fire support data.
- The Combat Service Support Control System allows brigade logisticians to automatically track support requirements and proactively push support to combat elements. It tracks fuel, weapons, maintenance, and other support measures for each vehicle in the brigade.[39]

These applications are bound together into a tactical internet by voice, video, and data networks, including Single-Channel Ground and Airborne System radios, the Enhanced Position-Location Reporting System digital transmission systems, and in the future, the Near-Term Digital Radio System.[40] The wider Army and joint network is built on an emerging infrastructure of MILSTAR, Defense Satellite Communications System (DSCS), and Global Broadcast System (GBS), satellite systems connected to the SMART-T (secure, mobile antijam, reliable, tactical terminal) and Trojan Spirit communications nodes.[41]

---

[39] *Army Transformation Taking Shape: The Interim Brigade Combat Team*, U.S. Army Center for Lessons Learned, Newsletter 1-18, 2001, Appendix A: Army Battle Command Systems Descriptions, available at http://call.army.mil/products/newsltrs/01-18/.

[40] Headquarters, Department of the Army, 2003d, Chap. 2, Sect. 6.

[41] U.S. Army Training and Doctrine Command, 2001b, Chap. 4, pp. 14–15. It is worth noting that the Army had immense difficulty making the tactical internet work during Operation Iraqi Freedom, indicating that further development and training are required. See "Army Must Overhaul Commo," 2003.

Looking further ahead, the Army's premier acquisition program for the next decade is the Future Combat System. This program began in 2000 as a partnership between the Army and the Defense Advanced Research Projects Agency. According to Army planning documents,

> The Future Combat System is comprised of a family of advanced, networked air- and ground-based maneuver, maneuver support, and sustainment systems that will include manned and unmanned platforms. The Future Combat System is networked via a C4ISR [command, control, communications, computers, intelligence, surveillance, and reconnaissance] architecture, including networked communications, sensors, battle command systems, training and both manned and unmanned reconnaissance and surveillance capabilities that will enable improved situational understanding and operations at a level of synchronization heretofore unavailable.[42]

In 2001, the Army assumed sole responsibility for the program, and in March of the following year, it selected a team formed by the Boeing Company and Science Applications International Corporation to be lead system integrator through the system design and development phase. In March 2003, DoD gave Milestone B approval to the program.[43]

The Future Combat System is intended to comprise 18 variants and the network that connects them.[44] These variants include an infantry combat vehicle, a mounted combat system, a non-line-of-sight cannon, a non-line-of-sight mortar vehicle, a reconnaissance and surveillance vehicle, a recovery and maintenance vehicle, a command-

---

[42] Headquarters, Department of the Army, 2003a, p. 47. See also Headquarters, Department of the Army, 2004.

[43] "Army Future Combat Systems Passes Major Milestone," Washington, DC: U.S. Army Public Affairs, available at http://www4.army.mil/ocpa/read.php?story_id_key=178. According to the National Defense Authorization Act for Fiscal Year 2003 (Section 8.8, Paragraph (b)(8)), "Milestone B approval" is a decision to enter into system development and demonstration pursuant to guidance prescribed by the Secretary of Defense.

[44] U.S. General Accounting Office, 2003a, p. 15.

and-control vehicle, and a medical vehicle.[45] According to Army plans, it will also have an armed robotic vehicle and four types of UAVs.[46] The Army envisions equipping each of its new brigade-level Units of Action with nearly 700 Future Combat Systems, although for the time being they are equipped with legacy vehicles.[47]

Within the Unit of Action, the Army envisions developing a command-and-control network that will link every vehicle and dismounted soldier with every other vehicle, soldier, and headquarters element. The network is also intended to have seamless interfaces with joint and multinational elements on the battlefield via voice, data, and video communications.[48] Two key programs for this network are the Warfighter Information Network–Tactical and the Joint Tactical Radio System.[49]

The Army expects the battle command network to raise the combat power of its Units of Action exponentially. However, the Future Combat System concepts are highly ambitious. The system will be heavily reliant on a complex communications network, implying challenges in bandwidth and interoperability.[50] Other areas of technical challenges include its advanced armor, sensors, powerplant, and weapons.[51] The Army admits that the Future Combat System pushes technology to its limits, but it expects that aggressive risk-management techniques and spiral development processes will allow the first phase of the Future Combat System to be fielded in 2012.[52]

---

[45] Riggs, 2003.

[46] "US Army Foresees 6,000-plus UAVs for Future Combat Systems," 2003.

[47] U.S. Army Training and Doctrine Command, 2002, p. 26.

[48] Headquarters, Department of the Army, 2002c, p. i. See also "Testimony of Steven W. Boutelle," 2004.

[49] U.S. General Accounting Office, 2003a, p. 16.

[50] See, for example, Burger, 2003.

[51] See, for example, Krepinevich, 2002.

[52] Headquarters, Department of the Army, 2003b, pp. 26–29.

In the meantime, the Army plans to integrate some subsystems of the system into current equipment as they mature.[53]

New Combat Organizations

As it transforms, the Army is also introducing new combat organizations. The initial focus is on the brigade echelon, where two new types of brigade are currently being fielded, the Stryker and the Brigade Unit of Action (BUA).

Stryker brigades are intended to bridge the gap between the Army's light forces, which are easily deployed but not well protected or lethal, and its heavy forces, which are well protected and lethal but not easily deployed.[54] A Stryker brigade weighs an aggregate 15,000 short tons, compared with about 25,000 short tons for a heavy brigade.[55] Five Stryker brigades are being organized in the active force and one in the Army National Guard.

The Stryker brigade is organized around the brigade headquarters, three infantry battalions, a cavalry squadron, an air-defense company, an artillery battalion, an engineer company, and a brigade-support battalion. In addition, it may receive augmentation from division and corps as required for the mission.

Stryker brigades are optimized to operate semi-independently in smaller-scale contingencies.[56] They have some organic capabilities that most other brigades would have to receive from higher echelons. The Stryker brigade has its own cavalry squadron, with sophisticated sensors, UAVs, and equipment to detect nuclear, chemical, and biological weapons. Stryker brigades also have a field artillery battalion, an antitank company, an engineer company, a military intelligence company, and a signal company. Each of these capabilities is normally held at division level or higher and is provided to brigades as priorities allow. Because the Stryker brigade is expected to operate

---

[53] U.S. Army Training and Doctrine Command, 2003, p.1.

[54] U.S. Army Training and Doctrine Command, 2001b, Chap. 1.

[55] Peltz et al., 2003.

[56] Headquarters, Department of the Army, 2003d, Chap. 1, Sect. 1.

independently in many smaller-scale contingencies, these capabilities have been integrated into its structure. In addition, the Stryker brigades can accept augmentation in other areas as required by the mission.

In addition to the five active Stryker brigades, the Army plans to create 43 BUAs by the end of 2008. There are heavy and light versions of the BUA. According to preliminary Army plans, the heavy BUA will feature two combined-arms maneuver battalions, a reconnaissance squadron equipped with various UAVs and other surveillance equipment, a strike battalion equipped with self-propelled howitzers and associated target-acquisition equipment, and various support troops. The heavy BUA maneuver battalions will feature two armor companies equipped with M1 Abrams main battle tanks and two mechanized infantry companies equipped with M2 Bradley infantry fighting vehicles. The battalions will also have detachments of mortars, snipers, medical personnel, engineers, TACs, and various other support troops to allow the brigade to operate relatively autonomously.[57]

According to preliminary Army plans, the light BUA design will feature two infantry battalions, a reconnaissance squadron equipped with UAVs and other surveillance systems, a strike battalion equipped with towed field artillery, and various support units. The infantry battalions will also have mortars, snipers, TACs, and an assault platoon of wheeled vehicles armed with heavy crew-served weapons. Light BUAs are intended to replace current light infantry, mountain, air assault, and airborne units.[58]

Together, the Stryker brigades and the BUAs represent a wholesale reorganization of the Army's tactical units. Current brigades are being converted to the new designs, and additional brigades are being created out of division end-strength. Though the Army's reorganiza-

---

[57] Draft Army planning documents. See also Feikert, 2004, pp. 8–9. Much information is also available in press accounts. See Hutcheson, 2004a; Sheftick, 2004.

[58] Draft Army planning documents and Feikert, 2004, pp. 8–9. See also Hutcheson, 2004a; Escoto, 2004.

tion plans are likely to continue to evolve over time, the themes of modularity and brigade-centric operations are likely to endure.

New Doctrine

As it transforms, the Army is revising its doctrinal approach to combat operations. For Air Force planners, the changes to Army maneuver doctrine and fire support doctrine are likely to be the most important.

Army maneuver operations will be transformed as advanced information technology helps commanders and staff to tighten their decision cycles.[59] Stryker brigades and BUAs will be dispersed over more territory than current units are, as much as 10,000 square kilometers for the Stryker brigades and perhaps more for the BUAs.[60] Their speed and dispersion will make new Army forces more difficult to target. Emerging doctrine focuses on "synchronized, simultaneous, combined arms attacks" that occur more rapidly than the enemy can respond.[61] Through networking, Army commanders can coordinate their actions to achieve massive effects without having to mass their forces.

Transformation is also changing Army doctrine for fire support.[62] In current doctrine, fire support coordinators at battalion, brigade, division, and corps levels develop fire support plans, allocate and position field artillery assets, and establish the division of labor in the combat zone.[63] This approach to fire support planning is highly

---

[59] U.S. Army Training and Doctrine Command, 2001b, Chap. 1; 2002, pp. 54–55.

[60] This compares to roughly 500 square kilometers or less for current brigades. U.S. Army Training and Doctrine Command, 2001b, Chap. 4, p. 33; 2001c, Chap. 6, p. 144.

[61] U.S. Army Training and Doctrine Command, 2001b, Chap. 4, p. 25; Chap. 6, p. 141; 2003, p. 4.

[62] Army doctrine defines fire support as "the collective and coordinated use of land- and sea-based indirect fires, target acquisition (TA), armed aircraft, and other lethal and nonlethal systems against ground targets in support of the force commander's concept of operations" (Headquarters, Department of the Army, 1993, Chap. 1).

[63] Ibid.

centralized and deliberate.[64] Fire support control measures tend to be linear and inflexible, typically including unit boundaries, permissive measures (such as free-fire areas or special engagement zones), and restrictive measures (such as no-fire zones and special coordination areas).[65]

Army Transformation promises to refashion the Army's approach to fire support planning, partly by expanding the concept to incorporate all forms of indirect effect on the battlefield. In the words of one Army planning document, the new fire support doctrine of "effects-based fires" represents

> an emerging operational, organizational, and doctrinal evolution within the Army regarding the planning and employment of fires and effects. In the past, Army fires were platform and system oriented. . . . The development of precision munitions and better non-lethal capabilities, coupled with advances in range, communications, ISR [intelligence, surveillance, and reconnaissance], and improved capabilities for routine employment of non-organic and joint service assets, are collectively leading to an orientation on effects rather than the systems that deliver fires.[66]

The Stryker brigade features new organizational and staff arrangements for coordinating fire support. In place of the traditional fire support element (FSE), the new brigades will have a fires-effects coordination cell headed by an effects coordinator.[67] This new organization is designed to conduct real-time dynamic integration of organic brigade target-acquisition assets, assigned Army assets, joint sensors, organic fire support assets, and other providers of fire to in-

---

[64] For an overview of the fire support planning process, see Headquarters, Department of the Army, 2002b, Chap. 1. For views on the ponderousness of the planning process, see Cheek, 2003, p. 33; Swortz, 2002, p. 12; and comments by Army planners in Pengelley, 2003, p. 26.

[65] Simpson, 2003, pp. 29–30; McDaniel, 2001. Army after-action reviews confirm the service's preference for linear fire support control measures ("Fires in the Close Fight: OIF Lessons Learned," briefing by Division Artillery, Third Infantry Division (Mechanized)).

[66] U.S. Army Training and Doctrine Command, 2001b, Chap. 3, p. 42.

[67] Ibid., p. 43.

clude large numbers of organic UAVs and unmanned ground vehicles. To meet these expanded requirements, the fire support communications network has been digitized and expanded via introduction of the Advanced Field Artillery Tactical Data System, which automates processing of targets and requests in accordance with the maneuver commander's targeting guidance.

Trends in Army Firepower

As the Army transforms, individual changes in fire support doctrine, organization, and equipment may be more or less important in their own right. Taken together, however, their interaction has important implications for Air Force counterland operations.

To understand the implications of transformation on Army fire support, the project team quantified the amount of potential fire support capability available to brigade-sized Army combat units during a notional tactical engagement. This approach was applied to a sample of more than 40 maneuver brigades deployed in operations from Operation Urgent Fury in Grenada (1983) to Operation Iraqi Freedom (2003). The methodology was also applied to the Army's future maneuver brigade designs, including the Stryker brigade and the BUA of the Army's Future Force. We compiled data for the following historical operations and generic cases:

Historical operations:

- Operation Urgent Fury, Grenada, 1983
- Operation Just Cause, Panama, 1989
- Operation Desert Storm, Southwest Asia, 1991
- Operation Restore Hope, Somalia, 1992–1993
- Operation Uphold Democracy, Haiti, 1994
- Operation Enduring Freedom, Afghanistan, 2002
- Operation Iraqi Freedom, Southwest Asia, 2003

Generic cases:

- Current Marine expeditionary unit
- Stryker brigade combat team
- Heavy BUA
- Light BUA

The methodology facilitated rough comparisons of firepower potential between brigades in the sample, defined as fire support available (measured largely as weight of fire and number of munitions) during a notional six-hour engagement. The methodology consists of the five steps described below.

First, the project team developed an understanding of the order of battle for U.S. land forces in the historical operations, using the best sources in the public domain. For the small-scale contingencies of the 1980s and 1990s, the team generally used data maintained by the U.S. Army Center for Military History.[68] For Operation Desert Storm, it used a book-length ground order of battle published commercially.[69] For recent operations in Afghanistan and Iraq, it used news reports.[70] And for the Stryker brigade combat team and the BUA, it used Army planning documents.[71] In each case, the team sought to develop an order of battle that was as detailed as possible at the brigade, division, and corps levels.

Second, the team estimated the number and type of weapons available to each maneuver brigade in the historical and generic cases, both weapons organic to the brigade and weapons habitually associated with the brigade, such as a field artillery battalion in direct support. Doctrinal manuals provide these data for each type of Army brigade and for a Marine Expeditionary Unit.[72] Doctrine normally specifies the number and types of indirect-fire weapons, typically mortars and cannon artillery, that a brigade controls in combat. The team assumed that a brigade would have doctrinal fire support arrangements unless historical data indicated that some other arrange-

---

[68] The Army Center for Military History provides order of battle at http://www.army.mil/CMH-pg/matrix/Matrix.htm.

[69] See Dinackus, 2000.

[70] Kraft, 2002; Cordesman, http://www.csis.org/features/iraq_instantlessons.pdf; see also "Iraq – US Forces Order of Battle," GlobalSecurity.org at http://www.globalsecurity.org/military/ops/iraq_orbat.htm.

[71] Headquarters, Department of the Army, 2001d. The staff of the Field Artillery Center at Fort Sill, Oklahoma, also provided invaluable assistance on this subject.

[72] Headquarters, Department of the Army, 1989, 1990a,b, 2000b.

ment pertained during a given operation. Data for the Stryker brigade combat team and BUAs were derived from Army planning documents.

Third, the team estimated the number and type of weapons from higher echelons that were available in historical cases or would likely be available in future operations, typically divisional and corps-level field artillery and attack helicopters. The primary sources for these data were historical orders of battle and Army field manuals for the field artillery brigade, divisional artillery, corps artillery, attack aviation battalion, aviation brigade, and division air cavalry squadrons.[73] The team assumed that each maneuver brigade in the sample had a proportional share of these fire support assets.[74]

Fourth, the team calculated fire support potential for each type of fire support asset, using several alternative methods. For example, Army documents provide "sustained firing rate" planning factors for each type of indirect-fire system.[75] Using these factors, it would be possible to calculate the number and weight of munitions a particular weapon could expend by multiplying the sustained firing rate by six hours, the duration of the notional engagement. But this would have vastly exaggerated the amount of ammunition that would be available to artillery and mortar crews in an actual situation. Instead, the project team used ammunition basic load planning factors to determine how much ammunition of what types would typically be available to forward units and in the supply system for immediate use.[76] Units

---

[73] Headquarters, Department of the Army, 2001c, 2003c,d.

[74] We realized that this may introduce some upward bias in our firepower estimates, as division and corps assets are often used to prosecute the deep battle rather than to support maneuver brigades, but as this upward bias would be uniform across the cases, we judged that it would not skew the elements of the analysis.

[75] "Army Fact File," Headquarters, Department of the Army, Washington, DC, available at http://www.army.mil/fact_files_site/fieldart.html.

[76] Headquarters, Department of the Army, 1990a. See Table 2-20, Ammunition Basic Load Guide, and Table 2-16, Ammunition Type Unit per Weapon per Day Expressed in Rounds and STON [short ton, i.e., 2,000 lb]. Headquarters, Department of the Army, 1990a, does not include basic load estimates for some recently fielded U.S. Army systems. In the present study, it was assumed that basic loads for new systems would be similar to basic loads for the systems they replaced.

were assumed to have expended half of their ammunition basic load during a six-hour engagement, reserving the balance for future engagements. For Army aviation, the team assumed that each helicopter flew one sortie during the six-hour engagement and expended its entire munitions payload.

The final step was simply to calculate the resulting fire support potential for each brigade in the sample. For example, the 1st Brigade of the 82nd Airborne Division air-landed at Panama's Tocumen Airport on the night of December 20, 1989, during the initial stages of Operation Just Cause. The organic fire support assets of the 1st Brigade of the 82nd Airborne Division included six 60-mm mortars, four 81-mm mortars, and four 4.2-in. mortars in each of its three maneuver battalions. In addition, the brigade was augmented by Battery A of the 319th Field Artillery for the assault (four 105-mm towed howitzers). The XVIIIth Airborne Corps, commanding the operation, retained two battalions of OH-58 Kiowa scout helicopters as a theater asset. The team assumed, for purposes of analysis, that the fire support potential of these helicopters would have been available to support maneuver brigades, if needed, on a relatively equitable basis. The fire support potential of the 1st Brigade of the 82nd Airborne Division in Operation Just Cause was thus estimated to be half the ammunition basic load of the available assets: 18 60-mm mortars, 12 81-mm mortars, 12 4.2-in. mortars, four 105-mm howitzers, and the full weapons payload of nine OH-58 helicopters.

A comparison with the 1st Brigade of the 1st Infantry Division in Operation Desert Storm is instructive. The 1st Infantry Division had a difficult mission on the first day of the ground war, breeching the border berm and directly assaulting an Iraqi infantry division on the other side of the border. In addition to the eight 81-mm mortars and eight 4.2-in. mortars in the infantry battalions, the brigade had its habitually associated field artillery battalion in direct support (24 155-mm self-propelled howitzers). The division also had been augmented by no less than five brigades of extra field artillery, comprising 203-mm howitzers, 155-mm howitzers, and Multiple Launch Rocket System (MLRS) batteries. The division also had the 4th Aviation Brigade's attack and scout helicopters in direct support and a share of the four field artillery brigades and the additional aviation brigade augmenting VII Corps. As a result, the fire support potential for the 1st Brigade of the 1st Infantry Division in Operation Desert Storm dwarfed that of the 1st Brigade of the 82nd Airborne Division in Operation Just Cause by nearly a factor of 50.

Finally, the Stryker brigades provide an example of the methodology for a prospective case. Each Stryker brigade fields 18 120-mm mortars in its infantry battalions. Additionally, each has an organic field artillery battalion equipped with 18 155-mm towed howitzers. However, because the brigades are intended to operate independently, the team did not assume that they would be provided division- and corps-level support when calculating their fire support potential. Additionally, Stryker brigades carry less ammunition than conventional brigades do in order to improve strategic deployability. Not surprisingly, the fire support estimate for a Stryker brigade is therefore somewhat greater than that for the 1st Brigade of the 82nd Airborne Division in Panama but only a small fraction of that for the 1st Brigade of the 1st Infantry Division in Operation Desert Storm.

Results

The project team focused on two measures of fire support potential during a six-hour engagement: number of munitions and tonnage of munitions. The resulting estimates for number of munitions are displayed in Figure 4.1.

The historical data suggest, not surprisingly, that units tend to have greater fire support potential when they participate in major combat operations, such as Operation Desert Storm, than they do in smaller contingencies, such as Operation Enduring Freedom.[77] But even in major combat operations, fire support potential appears to be declining over time. Brigades in Operation Iraqi Freedom had fewer

---

[77] Nevertheless, there are outliers, such as the 2nd Brigade of the 10th Mountain Division deployed to Operation Restore Hope (Somalia).

Figure 4.1
Fire Support Potential, by Number of Munitions

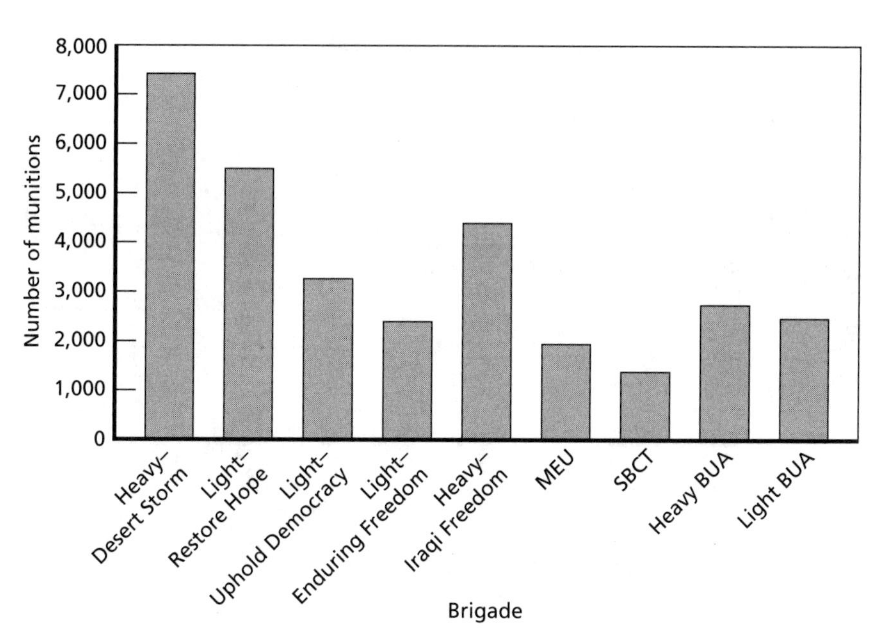

NOTE: Each bar represents one brigade from the sample. Brigades are identified by type (heavy, light, Stryker, etc.) and historical operation, if applicable. Heavy = armor or mechanized brigade, Light = light infantry brigade, MEU = Marine Expeditionary Unit, SBCT = Stryker Brigade Combat Team, Heavy BUA = Heavy Brigade Unit of Action, Light BUA = Light Brigade Unit of Action.

RAND *MG301-4.1*

munitions available than brigades in Operation Desert Storm did, and heavy BUAs will have fewer still.[78]

Estimates for tonnage of munitions are displayed in Figure 4.2. The outcome differs slightly from that in Figure 4.1 because munitions have different weights. For example, airborne and light infantry brigades have more mortars than do mechanized and armored bri-

[78] However, allocation of artillery may have been too low. The 3rd Infantry Division (Mechanized) reached this assessment: "A reinforcing artillery brigade with at least one cannon and one rocket battalion along with its associated counterfire radars is critical to support division offensive operations" (Headquarters, 3rd Infantry Division (Mechanized), 2003).

Figure 4.2
Fire Support Potential, by Tonnage

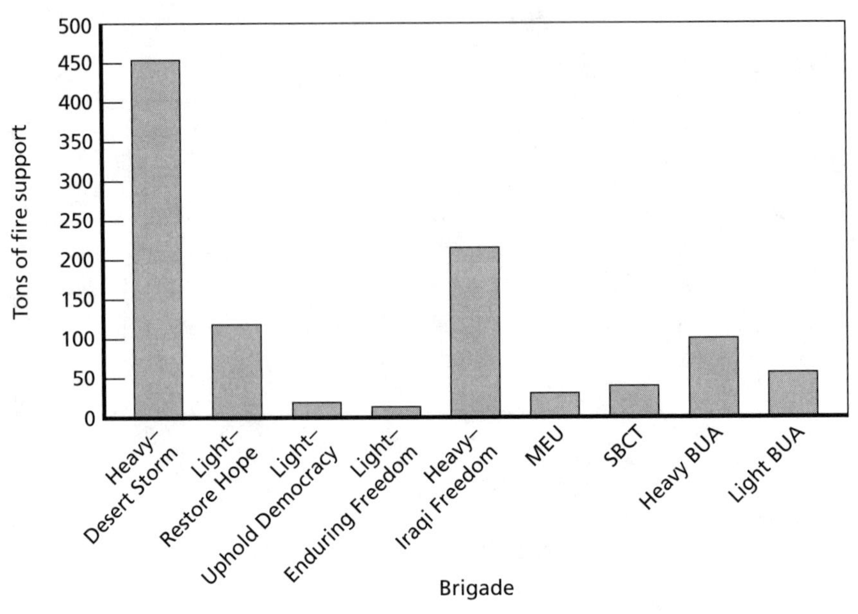

RAND *MG301-4.2*

gades, and they therefore fire more mortar rounds, but mortar rounds are lighter and less destructive than those fired by cannon artillery.

Some of the previously observed trends are still apparent when this different measure is used. Most important, the weight of fire support allocated to brigades participating in major combat operations appears to be declining over time. Heavy brigades participating in Operation Iraqi Freedom were allocated only half the fire support potential (measured in tons) allocated to heavy brigades in Operation Desert Storm, and Army planning documents indicate that allocations for heavy BUAs will be reduced by an equivalent measure. Our estimates therefore suggest that a heavy BUA will have only 25 percent of the fire support provided to heavy brigades in 1991. This 75 percent reduction will be Army Transformation's most important consequence, particularly for the Air Force.

Fire support estimates for the Stryker brigades and light BUAs offer a slightly different comparison with historical brigades. When fire support potential is measured by the number of munitions available, the new brigades appear to have less than historical brigades had. By weight of fires, however, they are comparable to historical brigades. This is likely due to the replacement of small mortars (60-mm, particularly) with the larger 120-mm mortar in the new brigade designs. However, the net effect is still likely to be a reduction in potential fire support for these brigades. Emerging doctrine suggests that the new brigades will be spread over areas of operation up to ten times as large as those of current light brigades, so that any given engagement is likely to occur outside the range of a significant proportion of the brigade's fire support assets. Our estimates are therefore probably too high. Furthermore, if the logistics and maneuver concepts for the new light BUA parallel those of the Stryker brigade combat team (as seems likely), our fire support estimate significantly overestimates the actual fire support potential of those brigades.

Broadly speaking, then, our analysis indicates that future brigades will have less fire support potential, measured in terms of both the number of munitions and the weight of fire, than historical brigades had. In the case of heavy brigades conducting major combat operations, the decrement is particularly significant both in numerical terms and in implications for Air Force counterland doctrine.

What accounts for these results? Two mutually reinforcing trends appear to be at work. First, the Army has adopted a brigade-centric concept of operations that implies that future brigades will operate independently and therefore without much of the division- and corps-level fire support provided in the past. Thus, the division- and corps-level systems that provided much of the fire support potential for historical cases (except Iraqi Freedom, where fire support capabilities were limited in line with emerging doctrine) are not present in the estimates for the Stryker brigade, heavy BUA, and light BUA.[79]

---

[79] It might be argued that the Army would in fact deploy these augmenting fire support capabilities in operations where they might be needed. This is certainly a fair point, and it suggests that our results should not be construed as a prediction regarding the fire support assets the Army will deploy to any particular future operation. However, in our view, it also underestimates the pressures upon the Army to reduce footprint in future theaters, the difficulty Army planners face in predicting the fire support needs in the early days of an opera-

Second, the deployability and footprint considerations that are driving the brigade-centric concept have also led the Army to reduce ammunition basic loads for future brigades, especially the Stryker brigades. For example, the planning factors for Stryker brigade 155-mm ammunition are about one-third less than historical planning factors, with consequent effects on fire support potential. The estimates for the BUAs have been calculated using standard logistics assumptions. If Army combat-service-support transformation leads to Stryker-like sustainment concepts for the BUAs, actual fire support for these brigades could be much lower than the estimates presented here.

It should be noted that the Army is not making these changes in a vacuum or unthinkingly. As discussed at the beginning of this chapter, the service's transformation is a response to a number of long-term trends in the nature of warfare. The Army is, for example, investing heavily in guided 155-mm artillery munitions, precision mortar rounds, and a variety of other precision fire support systems, in addition to upgrading reconnaissance, surveillance, and target-acquisition systems. Because our analysis is limited to gross measures such as weight and number of munitions, it might reasonably be argued that we underestimate the potential for precision firepower to provide equivalent battlefield effects with fewer and lighter munitions. There are several analytical reasons for reserving judgment on the effects of the precision munitions being developed by the Army.

First, there are clearly important fire support missions that will continue to require mass and numbers, especially when the effect must last over a period of time. An example might be a 30-minute suppression of an enemy ground unit, which would require a significant weight of fire, regardless of accuracy.

Second, while the Army plans to field a number of precision fire support systems, all of its existing programs are some years from be-

---

tion, and the Army's recent conversion of many field artillery units to other specialties—all of which militate against assuming that divisional and corps artillery will be arrayed on most future battlefields. The trend is illustrated by the dearth of division- and corps-level artillery in Operation Enduring Freedom and Operation Iraqi Freedom.

ing deployed.[80] By contrast, the new brigade designs are being implemented now. As this report is being written, the first Stryker brigade is in combat and the first set of BUAs is being formed. Therefore, even under the best assumptions, there will be some period during which the Army's new brigades will be fighting without the benefit of organic precision fire support.

Third, the Army has experienced great difficulty in developing and fielding precision fire support capabilities. Numerous programs have been canceled over the years, including the Enhanced Fiber-Optic Guided Missile, the Sense and Destroy Armor howitzer round, the Army Tactical Missile System Block II, and Brilliant Anti-Armor Technology. Furthermore, some of the precision fire support systems the Army has fielded, e.g., the Copperhead laser-guided 155-mm howitzer round, produced disappointing results in actual operations. Considering this record, it would be imprudent to assume that the Army's plans for precision-guided munitions will reach their full potential on schedule.

Implications for the Air Force

In some ways, Army Transformation will complicate Air Force counterland operations. Current coordination measures that rely on geographic boundaries (for example, the FSCL) will become less useful as ground operations evolve toward dispersed and nonlinear maneuver. Elements of Stryker brigades and BUAs may maneuver throughout future combat zones intermixed with adversary ground elements, making CAS more difficult. Army fire support operations will also be much more diffuse, dynamic, and difficult to predict in advance. All will be moving quickly to exploit opportunities and avoid threats, making deconfliction more challenging.

Command and control of counterland operations will also become more challenging. The current process of planning and allocating sorties works reasonably well, with the level of effort set through the air tasking order cycle and control handled dynamically

---

[80] Headquarters, Department of the Army, 2002a, p. 14.

through the ASOC or another battle-management system. But as the Army compresses its decision cycle, the current system may seem insufficiently responsive by comparison. Faster decision cycles on the ground will likely require faster decision cycles for air operations.

Future Army forces will rely more on air power to help them survive and to apply lethal firepower. The Stryker and Future Combat System are designed for rapid strategic deployment and therefore lack passive protection equivalent to that of today's Abrams and Bradley vehicles. Future Army forces will be operating in smaller elements that lack the mass and resilience of current forces. As a result, they might need highly responsive air power to deal with sudden emergencies. Air Force reconnaissance and surveillance capabilities may also play an important role in augmenting the situational awareness that emerging Army doctrine describes as key to survivability for transformed Army forces. Moreover, as has been demonstrated, the Stryker brigades and BUAs will not have as much organic fire support as Army brigades have traditionally received. As a result, air power might be a crucial source of additional lethality.

As a corollary, however, land forces will become more promising partners for air power. Adversaries usually adapt quickly to U.S. air attack by employing dispersal, concealment, and ambush tactics to minimize their vulnerability. Army Transformation is designed to produce forces to counter these tactics. Dispersed, nonlinear operations should help in gaining intelligence on enemy forces, flushing those forces, and compelling them to accept combat. In these ways, Army forces can expose adversaries to air attack and thereby serve a crucial role in Air Force counterland operations.

To improve its counterland capability, the Air Force should seize the opportunity afforded by Army Transformation. It should take a proactive role in developing new concepts and doctrine for air-land operations with new Army forces. The Air Force should also work to improve the links between Air Force and Army forces at all echelons. It needs to overcome aversion to close air attack merely because that operation accords land-force commanders control over air forces. At the same time, the Army needs to think about air power less as a supporting arm and more as a partner. Ground maneuver and air attack should be seen as mutually enabling elements in a joint air-ground team.

The Air Force and the Army need to develop mutual trust through routine training at the tactical level. CAS will never reach its full potential if the Army and the Air Force are strangers who meet on the battlefield. Soldiers should routinely incorporate CAS in their tactical plans, not as a substitute for artillery, but as a dynamically different source of fire. Airmen should recognize the unique demands of CAS and accord it a central place in their training. Separate training regimes tend to cause misunderstanding and, finally, a lack of trust. Only by training together at the tactical level can soldiers experience the enormous advantage of having friends in the air and can airmen grasp how best to help their friends on the ground.

CHAPTER FIVE

Air Attacks on Call

Introduction

Army Transformation envisions quick and decisive military operations conducted by forces that can be easily deployed. To make its forces lean, the Army intends to reduce fire support (mortars, artillery, and rockets) below historical norms. With less fire support in its own formations, the Army will tend to call for air attacks more frequently. Better situational awareness and more rapid maneuver may reduce the overall need for fire support, but there will still be many times when it is needed. Marines routinely use Marine air power to perform missions that Army soldiers would normally think more suitable for artillery and rockets. In the future, Army soldiers may begin thinking of air power as Marines do, with the important difference that they must rely on the Air Force, Navy, and Marine Corps to provide most of it. Moreover, air power will be critical to success or even to the survival of Army formations in more instances. The Army is currently fielding Stryker brigades with medium-weight armored vehicles and is developing a Future Combat System in the same weight class. These vehicles will have protection against rocket-propelled grenades and machine guns, but not against larger antitank missiles and cannons. Although they will still be able to close with an enemy, they will more often prefer to stand off and employ longer-range weapons or call for air attacks. Moreover, only air-delivered munitions may have the destructive effect and precision required for some missions, such as the destruction of an enemy strong point in urban terrain.

In most situations involving land combat, timeliness is a central issue. Once combat ensues, friendly land forces—even today's heavy forces—will seldom be in situations where timeliness is not critical. If a friendly force comes under mortar or artillery fire, it must bring counterbattery fire within minutes to prevent friendly casualties and to kill the enemy gunners before they displace. The standard during Operation Iraqi Freedom was two minutes from initiation of a fire request to firing the first rounds. Timeliness is so crucial in counterbattery fire that ground systems will usually be preferred to aerial systems for this mission. If a friendly force is ambushed, as has occurred countless times in Iraq, it must respond very quickly. Within seconds, it responds with its own direct-fire weapons, and they may be adequate to master the situation. If they are not, the friendly force will need additional fires within minutes to kill the enemy before he does more harm or slips away. In an extreme situation, air power may have to help protect a friendly force that is in danger of being overrun. This requirement is hardly new.[1]

Engagements occur at unpredictable times, implying that aircraft have to be on call for protracted periods. They might be already airborne or on strip alert, depending on the response time required. This chapter presents our analysis of aircraft on call. First, we identify the desired characteristics for aircraft performing attacks on call. Second, we focus on two of these characteristics that drive the number of aircraft required: the required response time and the amount of ordnance required per engagement. Third, we present a methodology to identify the total force structure required to support given levels of on-call fire. Finally, we use this methodology to determine the number of aircraft required, by type, to accomplish some representative tasks.

---

[1] One historical example is the action of the 1st Battalion, 7th Cavalry, in the Ia Drang Valley on November14–15, 1965.

Desired Characteristics for Aircraft on Call

The relative importance of the characteristics that are desirable for an aircraft tasked to conduct attacks on call depends on the situation. Any characteristic might be important or even critical at a given time. The first two characteristics listed below drive our analysis of requirements to keep aircraft on call. The others follow in random order:

- High airspeed
- Large weapons load
- Day and night, adverse-weather operations
- Long loiter time
- Situational awareness
- Quick turn (revisit) rate
- Mixed weapons load
- Accurate weapons delivery
- Survivability against air defense
- Flexibility to operate from unimproved bases

High airspeed allows an aircraft to arrive more quickly in the target area. It allows fewer aircraft to meet a requirement in terms of area coverage and time to respond, but once an aircraft arrives in the target area, slower speed and tighter turn become advantages. Moreover, high airspeed normally implies less loiter time.

Large weapons loads allow aircraft to handle more targets and thus to provide greater coverage. Larger loads lessen the chance that an aircraft will exhaust its ordnance before the engagement on the ground has ended.

Aircraft flying CAS should be capable of operating day and night under most weather conditions. Most U.S. fixed-wing attack aircraft have or are acquiring this capability.

Long loiter time confers several important advantages. It allows aircraft to remain longer on station, implying that fewer aircraft are required to maintain coverage. More important, it allows a pilot to continue his attacks until the targets are destroyed or the engagement has concluded. Otherwise, the mission might have to be handed off to a new pilot, who would initially lack the first pilot's understanding of the situation. There is an obvious relationship between loiter time and situational awareness.

Situational awareness varies widely from one aircraft to another. AC-130 crews, for example, acquire so much data through their sensors that they can perform valuable reconnaissance for land forces. A-10 pilots often acquaint themselves with the intentions of friendly land-force commanders before they fly their missions. They perform reconnaissance simply by looking through the canopy at low altitude, sometimes a risky maneuver. Pilots flying at higher altitudes can gain situational awareness through modern targeting pods at much less risk, but their field of vision is highly restricted.

Aircraft that employ free-fall weapons and forward-firing ordnance must make passes at their targets. If an engagement requires more than one pass, the ability of the aircraft to revisit can be critically important. During the first hours of Operation Anaconda, for example, bombers took longer than fighters to revisit targets. In contrast, an AC-130 aircraft orbits over its targets and does not need to make multiple passes. Future attack aircraft may have the ability to engage targets from any azimuth.

Carrying a mixed weapons load allows pilots to be effective against a variety of targets. When attacking targets on call, pilots will seldom know in advance what type of target they will be attacking. Carrying a mixed weapons load hedges against this uncertainty. Some targets require great precision, while others demand area coverage. Accuracy may be highly important to prevent fratricide and to limit the risk of collateral damage. Large warheads may be required in some situations and may be unacceptable in others, due again to collateral damage. Small warheads will often be required to limit collateral damage and reduce the risk of fratricide.

In recent operations, U.S. aircraft have enjoyed near impunity flying at medium to high altitude, above the reach of ubiquitous low-level threats ranging from man-portable air-defense missiles to cannons. Flying against more-sophisticated radar-guided air-defense missiles, U.S. forces would have to employ stealthy aircraft and standoff weapons. A-10 aircraft often fly within range of low-level air defenses, but their pilots survive by limiting their exposure and varying their approach, and by the toughness of the aircraft. The AC-130 gunship cannot fly above such air defenses without becoming ineffective due to the short ranges of its weapons. As a result, the AC-130 usually has to confine its operations to periods of darkness.

The flexibility to fly from unimproved bases allows aircraft to achieve longer loiter times with less recourse to aerial refueling, thus potentially reducing the number of aircraft required to achieve coverage. In addition, such flexibility allows aircraft to be poised on strip alert in locations near remote areas where friendly troops may be operating. Although onerous and taxing, strip alert imposes much less burden than does flying in orbit while awaiting calls, especially if calls are relatively infrequent. Of course, forward basing depends critically on the ability to secure those bases and to provide maintenance.

In addition to these factors, crew training has a great effect on performance. Every attack aircraft has some capability to perform CAS, but realizing that capability requires training. Training is often constrained by competing demands on scarce resources. For example, aircrews that have to maintain proficiency in suppression of CAS may have to scant the CAS mission. We omitted training from our calculations because it defied simple quantification, but it is highly important.

Assessing Required Aircraft

The following analysis addresses the problem of having aircraft available on call to support land forces. Of the desirable characteristics listed above, it is concerned with only the first two: high airspeed and large weapons load. It does not assess the relative capabilities of aircraft to perform these missions, beyond their ability to be continuously available and to deliver large weapons loads.

To assess how many aircraft would be required to respond on call, we considered a case in which land forces were operating within a box-shaped area (for example, 100 kilometers on a side). We postulated that land forces operating anywhere in this area would make random calls for fire. For fighters, we postulated that one two-ship formation (lead and wing) would be flying randomly within a "stack," i.e., airspace where aircraft fly in a holding pattern while awaiting calls, at the center of this area. For bombers and the MQ-9 Predator B,[2] we assumed one aircraft on station. Knowing the speed of an aircraft, we computed the time required for it to fly from the "stack" to the location of the friendly ground force. We added a time delay that accounted for command-and-control procedures, for TACs guiding the pilots to the target (talk-on), and for weapons delivery. We ran the calculation numerous times using a Monte Carlo model to determine the distribution of times within which the aircraft could respond. We set the desired response level at the 90th percentile of this time distribution (i.e., the aircraft could respond within that time in 90 of 100 random calls for close support). Knowing the area that the aircraft could cover and the munitions required per engagement, we calculated the number of aircraft to assure coverage.

When aircraft are responding to a call, they are unavailable for other calls. Therefore, each area covered would correspond to some land force that needed air support on call. Adding another land force with the same need would add another area and another corresponding flight. The number of land forces and corresponding areas could multiply rapidly, increasing the number of aircraft required to maintain coverage. However, it would be very inefficient to allocate air forces on a one-to-one basis in support of tactical units on the ground. Most sorties would be kept idle, waiting for missions that never came. Moreover, holding aircraft to support some unit without immediate need while denying aircraft to some unit with immediate need would not usually be justified. Therefore, aircraft are usually allocated to large formations, e.g., V Corps or the 3rd Infantry Division during Operation Iraqi Freedom, and they are released to those units with immediate need of air support. Except in the case of SOF, air support is seldom critical to the survival of land forces, which usu-

---

[2] The MQ-9 Predator B is a much larger aircraft than the preceding MQ-1 Predator A, capable of carrying up to ten AGM-114 Hellfire missiles.

ally can at least defend themselves until air support arrives. Endangered units normally receive the highest priority, for example, the Broken Arrow priority employed during the Vietnam War.

Force Structure for Protracted Coverage

This analysis assumes that close support would be required continuously, i.e., 24 hours a day, seven days a week. The force structure required to meet such a commitment has two components: number of aircraft and number of crews. We computed both numbers to gain a better understanding of the costs associated with a commitment. To identify the number of aircraft required, we drew on a simple sortie-rate model based on earlier unpublished RAND work.[3] This model, based on empirical data, determines the sortie rate according to the following formula:

$$SR = 24 / (FT + TAT + MT)$$

where

SR = sortie rate
FT = flight time
TAT = turnaround time
MT = maintenance time

Flight time includes all the time that aircraft remain airborne performing a mission, whether waiting in a "stack" or conducting attacks. Turnaround time is all the time required to prepare aircraft for their mission, including ordnance handling, checkout procedures, and taxiing to the runway. For this analysis, we assumed that turnaround time would be three hours for all aircraft types. Maintenance time is the time for regularly scheduled maintenance, calculated as

[3] J. Lawrence Hollett, unpublished RAND research on Air Force responses to the use of WMD, specifically, the standoff tactical air-power projection option. The associated sortie-rate model appears in Stillion and Orletsky, 1999, App. B.

3.4 hours plus 0.68 of the flight time. We assumed two hours for the maintenance time of the AH-64D.

At any given time, some aircraft will be unavailable due to breakdowns. To account for these aircraft, we applied the rates for mission-capable aircraft experienced during Operation Iraqi Freedom.[4]

The crew requirement is based on Air Force policy on flight-hour restrictions. The requirement could be reduced by using UAVs if they were flown in formations and each formation required only one crew to operate. Although these hour restrictions can be waived in particular cases, they provide good rules of thumb for determining pilot requirements. They would very likely apply to commitments to remain on call that lasted a month or longer. For commitments of such duration, the Air Force sets a tempo that is sustainable without adverse impacts on training, safety, and retention of personnel.[5] For the following analysis, we assumed that aircrews would be limited to 125 flight hours during 30 consecutive days. For commitments lasting longer than a month, the Air Force might impose a narrower constraint.

On Call During a Campaign

During a campaign, air forces may have to accomplish several tasks simultaneously, as they did during Operation Iraqi Freedom. In northern Iraq, U.S. light forces and SOF supported friendly Kurdish forces (*Peshmerga*) against Iraqi regular Army and Republican Guards

---

[4] See Department of the Air Force, 2003a.

[5] Department of the Air Force, Air Force Instruction 11-202 (Department of the Air Force, 2003b). AFI 11-202 offers several different sets of criteria. One set limits flying hours during a duty day, while another set limits flying hours during a specific number of calendar days. In terms of calendar days, flight hours are not to exceed 56 hours during 7 consecutive days, 125 hours during 30 consecutive days, and 330 hours during 90 consecutive days (p. 65). These criteria reflect the fact that crews can "surge" (fly longer) for a short period but cannot sustain this tempo for a long period. For this analysis, we assumed a limit of 125 flight hours during 30 consecutive days, the mid-level constraint. The requirements for crew rest will also constrain operations.

forces. In western Iraq, U.S. and Allied SOF searched for Iraqi ballistic missiles and eventually seized several airfields. In central Iraq, U.S. conventional forces, principally the 3rd Infantry Division (Mechanized) and the 1st Marine Division, conducted a rapid advance against remnants of Iraqi conventional forces and larger numbers of paramilitary Fedayeen forces. In southern Iraq, various Iraqi forces posed a threat to the U.S. line of communication extending from Kuwait to Baghdad, especially around An Nasiriyah and An Najaf.

Based on the experience in Iraq, we identified four representative tasks to gain a better understanding of the demand for on-call air power. Table 5.1 summarizes these tasks, describing each with regard to (1) the area within which the targets are located, (2) the time within which the attacks must occur, and (3) the required effort, measured as tonnage dropped on fixed targets or the number of mobile targets to be attacked.

For each of these tasks, we computed the number of aircraft required to respond to a call for five different types of aircraft: B-1B, F-16 Block 50, A-10, AH-64D, and MQ-9. The time to respond includes the command-and-control time, the time for the aircraft to fly to the target area, and the time to deliver one or more weapons (i.e., time for control procedures, time for talk-on in the target area, time for weapons to descend from the aircraft to the target, etc.). These times vary widely according to the situation. Our objective was to identify reasonable times for each task in order to assess relative availability of each aircraft type for the various tasks.

Table 5.1
Representative Tasks

Task	Area	Time	Effort
Breach line	50x50 km	1 hour	10 tons on fixed targets
Kill emerging targets	100x100 km	15 minutes	2 mobile targets
Halt enemy attack	50x50 km	30 minutes	20 mobile targets
Destroy strong point	100x200 km	1 hour	1 ton on a fixed target

We assumed two minutes for command-and-control time, i.e., the time from the call for fire until the aircraft are allocated and start moving toward the target area. We assumed different times for delivery of weapons after arrival in the target area. For the F-16, A-10, and AH-64D, we assumed that the first weapon could be delivered in two minutes and each subsequent weapon could be delivered in one minute. For the case in which these aircraft that typically operate in two-ship formations engage more than one target, we assumed that both aircraft can deliver weapons in the stated time. Because the situational awareness of the MQ-9 will not be as good as that of the piloted aircraft, we assumed four minutes for delivery of the first weapon and two minutes for delivery of subsequent weapons. The delivery time is a little different for the B-1B, since a weapons officer is on board, and we assume that all weapons are GPS-guided. We assume that it requires one minute for the weapons officer to deliver a single JDAM. In the case of mobile targets, we assume that a pattern of ten CBU-103s will be delivered and that it takes eight minutes for the weapons officer to input the coordinates. Subsequent attacks are assumed to take the same amount of time. Since a weapons-systems officer is on board the B-1, this process would start when the crew received the nine-line briefing and could thus be accomplished while the aircraft is en route to the target area. The eight minutes becomes a factor only when it is in excess of the time required for the B-1B to fly to the target area. For the other aircraft, these engagement times would start after the aircraft arrived in the target area.

As stated earlier, the times presented here are notional; in reality, there would be much variation, depending on the situation. For example, in some cases, an AH-64D Apache can engage a second target only seconds after it engages the first target if both targets are within range. In other cases, the time to reengage may be significantly longer if the AH-64D does not have line of sight or begins to take fire, forcing it to maneuver. As another example, the B-1B may require more time to engage additional targets in the same area, because it needs time to fly back. In part, the time to reengage is driven by the flight characteristics of the aircraft. For example, a B-1B flying at 450 knots needs 1.5 minutes to make a 360-degree turn while experiencing 1.5 times the normal force of gravity (1.5 g). By comparison, an A-10 flying at 300 knots takes about a half-minute to make the same turn at 3 g. Moreover, the A-10 pilot will normally come out of his turn ready to reengage, while the B-1B pilot may need to fly further until he reaches a suitable release point. However, the A-10 pilot may need to take several passes on the target before he is convinced that he can safely deliver ordnance in the target area, whereas the B-1B is launching on a coordinate. The times chosen for the particular functions were thus intended to be reasonable approximations of highly variable times.

To estimate how many targets each aircraft could attack, we identified alternate weapons loads against stationary and mobile targets for each type of aircraft. We assumed that the B-1B would employ 24 2,000-pound GBU-31 JDAMs to attack stationary targets and either 30 CBU-103s (CBU-87 combined-effects munitions equipped with the wind-corrected munitions-dispenser tail kit) or 30 CBU 105s (CBU-97 sensor-fuzed weapons with the same tail kit) to attack mobile targets. We assumed that these weapons would be employed at the rate of ten per engagement when attacking armored formations. Table 5.2 presents the weapons loads for the five aircraft considered. Since the B-1B and the F-16 were assumed to be operat-

Table 5.2
Weapons Loads

Target	B-1B	F-16 Block 50	A-10	AH-64D	MQ-9
Stationary	24 tons (12 GBU-31s)	2 tons (2 GBU-10s)	3 tons (12 Mk-82s)	0.93 ton (8 AGM-114Ks, 76 Hyrda-70s; 30-mm cannon)	1 ton (2 GBU-16s)
Mobile	30 targets (30 CBU-103s or 30 CBU-105s)	4 targets (4 CBU-103s or 4 CBU-105s or 4 GBU-12s)	10 targets (6 AGM-65s plus GAU-8 30-mm Gatling)	16 targets (16 AGM-114Ks plus M230 30-mm chain gun)	10 targets (10 AGM-114Ks)

NOTE: A = attack; AGM = air-ground missile; AH = attack helicopter; B = bomber; CBU = cluster-bomb unit; GAU = gun, automatic; GBU = guided bomb unit; MQ = modified drone.

ing from higher altitudes (about 20,000 feet above ground level), one minute was assumed for weapon flight time. For the other aircraft, we assumed weapon flight times in seconds and therefore neglected them in the calculations.

We used several additional parameters to complete the calculations (see Table 5.3), including airspeed for each aircraft, loiter time, transit time from the base to the target area, mission-capable rates, and turn time. True airspeed was used to calculate how quickly an aircraft could reach the target area. The transit time was used to compute the sortie rate for the aircraft and reflects the different distances from base depending on the type of aircraft. For example, B-1B bombers would usually be based farther from the target areas and would therefore have longer transit times. In contrast, AH-64D helicopters would usually be based close to the battlefield, perhaps in forward arming and refueling points, and therefore would have short transit times. The mission-capable rate, expressed as a decimal, reflects readiness typically experienced for each type of aircraft based on experience during Operation Iraqi Freedom.[6] The turn time is the

Table 5.3
Additional Parameters

Parameter	B-1B	F-16 Block 50	A-10	AH-64D	MQ-9
Speed (kt)[a]	470	470	380	143	270
Loiter time (hr)	8	4	3	3	24
Transit time (hr)	4	2	3	0	6
Mission-capable rate	0.794	0.739	0.85	0.682	0.766
Aircraft turn time (hr)	6	3	3	3	3

[a] These speeds were largely drawn from *Jane's All the World's Aircraft 2004–2005*. We assumed the A-10 speed on the basis of the information available and reduced it slightly to account for our larger weapons load. The B-1B and the F-16 are both supersonic platforms, for which we assumed a high subsonic cruise speed of 470 knots.

---

[6] See CENTAF, 2003, p. 10. The AH-1W was used to represent the AH-64D, since data for the AH-64D were not available.

time required to prepare an aircraft to fly after it returns from a mission.

Using the tasks presented in Table 5.1, the times for command and control, talk-on, and fly-out discussed earlier, the weapons loads presented in Table 5.2, and the additional parameters presented in Table 5.3, we calculated the numbers of "stacks" and the numbers of aircraft required for continuous coverage to respond within the required times. The time requirements presented in Table 5.1 are treated as 90th percentile. From this calculation, we determined the number of aircraft that would be required by evaluating the munitions loaded on each type of aircraft and the time required to engage multiple targets (if necessary). In some cases, only one aircraft per "stack" may be required. In other cases, more than one aircraft per "stack" may be required, because one aircraft may not carry enough weapons to engage all targets. When the number of targets exceeds the munitions carried by one aircraft, several aircraft attack sequentially from the same "stack." When the time exceeds that which can be achieved by one "stack," a second "stack" is added in a different geographic location (within the target area) to shorten the response time.

We also considered operational aspects. For example, F-16 and A-10 fighters normally fly in pairs (lead and wing), while B-1B bombers and MQ-9s normally fly singly. Finally, we used the sortie-rate model and the maximum-pilot-hours methodology discussed earlier in this section to compute the requirements for aircraft and aircrews to achieve 24-hour coverage. Table 5.4 presents the results. For example, to accomplish any of the four tasks using B-1B aircraft, five aircraft and nine aircrews orbiting in one "stack" would be required. Requirements for aircrews are based on Air Force planning factors for maximum flying hours per month.

The numbers of aircraft and "stacks" shown in Table 5.4 would be required to accomplish just the tasks within a target area, not to cover an entire country the size of Iraq. If several tasks had to be accomplished simultaneously, the numbers of aircraft and "stacks" would increase proportionately, although some economies of scale would apply.

Table 5.4
Aircraft, Aircrews, and "Stacks"

	Number of Aircraft/Number of Aircrews (Number of "Stacks")				
Task	**B-1B**	**F-16 Block 40**	**A-10**	**AH-64D**	**MQ-9**
Breach lines	5/9 (1)	34/52 (3)	26/47 (2)	—	—
Kill emerging target	5/9 (1)	12/18 (1)	13/24 (1)	16/24 (2)	4/8 (1)
Halt enemy attack	5/9 (1)	34/52 (3)	13/24 (1)	8/12 (1)	7/15(2)
Destroy strong point	5/9 (1)	12/18 (1)	13/24 (1)	8/12 (1)	4/8 (1)

NOTE: Numbers of aircraft and crews do not always scale with the number of "stacks," due to rounding. AH-64D and MQ-9 have no entries for "breach lines" because they are unsuited for the task.

This analysis shows that bombers are well suited to perform on-call missions. Indeed, one B-1B alone can carry enough payload to accomplish any of the tasks. Fighters, such as the F-16, are best suited for tasks that require quick response and rapid reengagement, not for tasks that require large amounts of munitions. Fighters might be made more effective against area targets by having them carry larger numbers of smaller munitions, such as the Small-Diameter Bomb. Because of its slow speed and limited weapons load, the MQ-9 is best suited to engage discrete targets when there is no requirement for rapid reengagement. The AH-64D and MQ-9 are unsuitable for tasks that require large amounts of munitions, e.g., breaching an enemy line. The A-10 and the AH-64D both excel in conducting rapid attacks on mobile targets.

If the required response time is fairly long, strip alert can be a useful technique. In strip alert, aircraft are staged on a runway apron, ready to take off when a call comes. Figure 5.1 shows the time to engage targets as a function of distance from the target area for combat air patrol and strip alert. This calculation assumes use of the F-16, a relatively fast and agile aircraft. If about 45 minutes are allowed to respond to a call, then strip alert becomes a viable option, assuming that a suitable base can be found within 200 kilometers of the target area. We used a fairly conservative assumption of 30 minutes for time to "wheels up," i.e., takeoff from the runway. If a few minutes could

Figure 5.1
Strip Alert vs. Combat Air Patrol

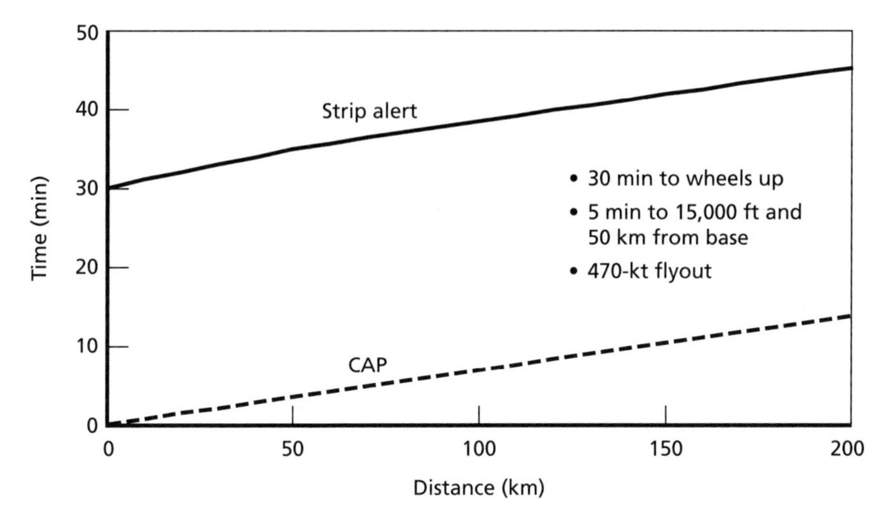

NOTE: CAP = combat air patrol, i.e., aircraft fly continuously above the area where they may have to strike targets.

RAND *MG301-5.1*

be shaved from this time, the response would be even better. We assumed that it takes five minutes to climb to 15,000 feet altitude and reach an area 50 kilometers from the base. From that point on we assumed 470-knot speed.

Strip alert requires fewer aircraft than does combat air patrol, as long as the number of scrambles, i.e., takeoffs in response to calls, remains modest. Figure 5.2 explores this relationship by computing the number of crews required to maintain two aircraft on strip alert as a function of scrambles for three average sortie durations ranging from two to four hours. Eight pilots are normally required to maintain a flight of two aircraft (lead and wing) on strip alert. This calculation assumes that each pilot will pull about 40 hours of strip alert time per week, or 548 alert hours per quarter. As the number of scrambles increases, the limiting factor becomes the maximum number of hours each pilot is permitted to fly over a specified period of

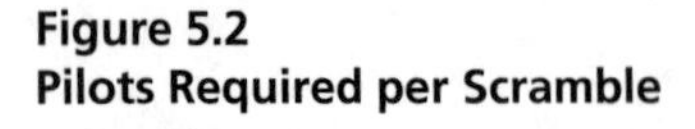

Figure 5.2
Pilots Required per Scramble

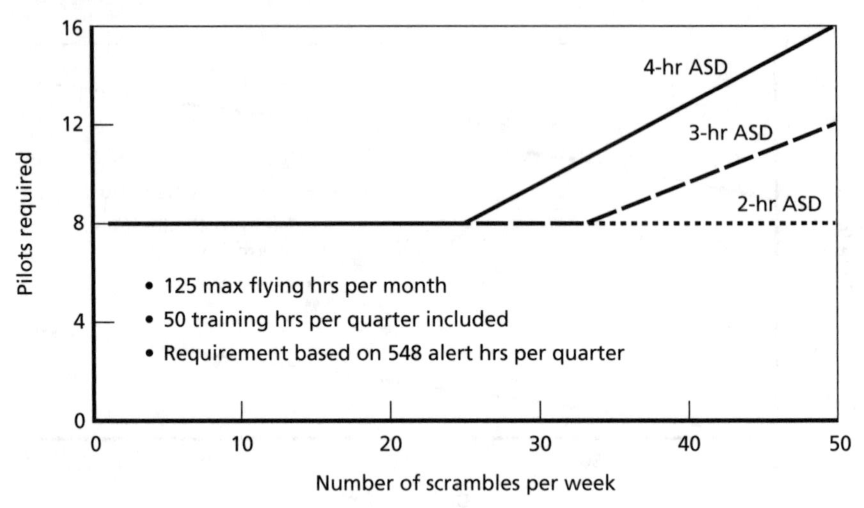

NOTE: ASD = average sortie duration.
RAND *MG301-5.2*

time in accordance with Air Force Instruction 11-202, Vol. 3. For our calculations, we assumed a constraint of no more than 125 flying hours per 30-day period, and we included 50 training hours per quarter.

Conclusion

These calculations suggest some insights into the feasibility of having aircraft continuously airborne to support land forces. A large manned bomber, such as the B-1B, is the most capable aircraft in this role. One continuously orbiting B-1B could accomplish any of the tasks examined in this study. However, to maintain this bomber on station would require a commitment of five aircraft and nine crews, unless flight hours were allowed to exceed prescribed limits.

It would be prohibitively expensive to have fighters continuously on station, ready to perform tasks that require large numbers of munitions. Even tasks that require smaller numbers of munitions, such as strikes on emerging targets and destroying enemy strong points, demand large numbers of fighters if they must be constantly available. One possible solution would be to use fleets of UAVs constantly orbiting in areas of interest.

In a large-scale operation, there are usually periods when demand can be anticipated, for example, when the lead brigade of the 3rd Infantry Division crossed the Euphrates River south of Baghdad, during Operation Iraqi Freedom. At such times, it makes sense to "stack" aircraft in anticipation of demand and to "push" them, i.e., keep aircraft flowing toward a target area until demand ceases. But too many aircraft and crews would be required to sustain continuous coverage indefinitely. In particular, trying to substitute aircraft for a land force's organic fire support (mortars, cannons, rocket launchers) would require too many aircraft and would forfeit the inherent flexibility of air power.

In a permissive air-defense environment, the B-1B has enormous potential as an "arsenal plane," due to its long range, high speed, large payload, and long loiter time over the target area. In this study, we assumed that B-1Bs would carry their current munitions, including GBU-31s, CBU-103s, and CBU-105s, and would receive targeting data from a TAC. The B-1B could become more lethal by adding a modern targeting pod and employing laser-guided weapons for greater precision. To improve situational awareness, the crew might also receive inputs from remote sensors. In a higher-threat environment, the next-generation gunship might play the role of "arsenal plane."

All aircraft could be more effective if they carried larger numbers of smaller bombs. Often the desired effect can be achieved with smaller warheads than those traditionally carried. In these cases, having larger numbers of bombs implies the ability to engage more targets. Smaller warheads can allow engagement closer to friendly forces without undue risk of fratricide and can also lower the risk of collateral damage to civilians.

CHAPTER SIX

Terminal Attack Control in the Air-Land Partnership

Introduction

Terminal attack control is the vital link between ground maneuver and airborne firepower. Whether airborne or on the ground, the TAC connects the ground commander to air power. The TAC will usually be the only person who understands both the ground-combat situation and how to best use combat aircraft to achieve the tactical objectives.

This chapter discusses the terminal attack control function, quantifies the growing demands for TACs, and presents some new concepts for meeting this demand.

Background

TACs were first used to direct air strikes in support of engaged ground forces during World War II.[1] During the Italian campaign, the Fifth Army conducted the "Rover Joe" experiment, in which an experienced fighter or bomber pilot stationed on high ground identi-

[1] Over the years, various terms have been used to describe terminal attack control. Initially, this mission was called "forward air control" and the person who conducted the mission was either a ground forward air controller (G-FAC) or an airborne forward air controller (A-FAC). In this report we use terminal attack controller (TAC) as a shorthand for all personnel—officer and enlisted, airborne or on the ground, from any service—who are trained and certified to control aircraft flying CAS.

fied targets for friendly aircraft.[2] Gen. Pete Quesada, Commander of the IX Tactical Air Command supporting the U.S. First Army, further refined this concept during the breakout from Normandy. Some Sherman tanks were equipped with Air Force radios, and the loader was replaced with a fighter pilot trained to do the loader function and act as a FAC. These special tanks were positioned near the front of attacking columns, allowing FACs to direct P-47 aircraft and other fighters against German forces impeding the Allied advance.[3] In the XIX Tactical Air Command (supporting Gen. Patton's Third Army), TACs rode in tanks and armored personnel carriers.[4]

During the Korean War, personnel of the 6147th Tactical Air Control Group flew as TACs in the T-6D Mosquito, a single-engine prop aircraft previously used as a trainer. A two-man team consisting of an Air Force pilot and an Army forward observer flew the missions. The Mosquitos created the first airborne forward air control system and the first TACP. In Korea, TACPs functioned primarily as communication links between airborne TACs and ground commanders. Many of these skills were lost when the Mosquitos were disbanded in 1956. A decade later, in Vietnam, the Air Force had to rebuild both the airborne TAC and TACP capabilities.[5]

During the Cold War, the Air Force and other services developed a comprehensive theater air control system and a variety of systems and procedures to provide CAS. The Air Force established the 1C4 Air Force Specialty Code for enlisted air command-and-control specialists; established the Air Ground Operations School to train air liaison officers, TACs, and airborne TACs; and created air support operations squadrons (ASOSs) and the ASOC. Each ASOS was associated with a specific Army unit, providing air liaison officers (ALOs) at the division, brigade, and battalion levels and ETACs down to se-

---

[2] Lester, 1997, pp. 10–11.

[3] Hughes, 1995, pp. 183–184.

[4] Spires, 2002, p. 74.

[5] See http://www.wpafb.af.milmuseum/air_power/ap55.htm (last accessed on November 6, 2003). See also Futrell, 1983, pp. 104–109.

lected companies. Air Force personnel were based with or near the supported Army unit to provide opportunities for regular interaction and training. The ASOC was assigned at corps level to coordinate counterland operations and allocate CAS resources. In the special operations world, the Air Force currently provides ALOs and ETACs for terminal attack control, as well as Air Force Special Operations Command (AFSOC) combat control teams (CCTs). CCTs provide combat air traffic control and terminal attack control for special missions. For example, a Ranger battalion conducting a parachute assault on an airfield would be accompanied by both its regularly assigned ETACs and ALOs for terminal attack control and CCTs to control the air drop, clear the airfield, set up navigation aids, and control subsequent air drops and aircraft landings. CCTs also deploy on missions in which their primary responsibility is terminal attack control. Examples of the latter include operations in Mogadishu in 1991 and Operation Enduring Freedom in Afghanistan in 2001.[6]

The Terminal Attack Control Mission

As noted above, terminal attack control may be accomplished by an airborne TAC, a ground-based ALO, an ETAC, or an AFSOC combat controller.[7] In every case, the TAC must be expert in the tactical

---

[6] Air Force Staff Sergeant Jeff Bray, a combat controller, was awarded the Silver Star for his actions directing Army AH-6 Little Bird attack helicopters during the "Bloody Sunday" battle in Mogadishu, Somalia, on October 3, 1993. See Bowden, 1999, p. 257; Oliveri, 1994; Tyson, 2002.

[7] We are discussing only Air Force personnel in this section. U.S. Marines also have airborne FACs, ALOs, and TACPs with similar capabilities. In addition to providing support for Marine aviation, Marine TACs have acted as liaisons for joint operations since World War II, when Marine assault signal companies coordinated naval CAS and gunfire in support of U.S. Army operations in the Pacific Theater. After World War II, the Marines created the Air-Naval Gunfire Liaison Company (ANGLICO) to provide this function. The current ANGLICO mission is "to support the U.S. Army or Allied Division, or elements thereof, by providing the control and liaison agencies associated with the ground elements, in the control and employment of Naval Surface Fire Support and Naval Close Air Support in the amphibious assault, or in other type operations when supported by Naval Surface Fire Sup-

application of air power and familiar with the capabilities and limitations of aircraft platforms (bombers, fighters, gunships, and attack helicopters), the characteristics of munitions (guns, missiles, rockets, gravity bombs, laser-guided bombs, and GPS-guided bombs), and delivery tactics. He must have a clear understanding of the ground situation to ensure that the aircraft and munition are appropriate for the mission, generating the desired effect against the enemy at acceptable risk to friendly forces. An experienced TAC can quickly determine if an aircraft flight path or dive angle is suitable, aborting the mission if the approach would threaten friendly forces. Even with advanced munitions such as laser-guided bombs (LGBs) or JDAMs, the aircraft approach vector is still important. Whenever possible, TACs will direct approaches such that weapons do not fly over friendly forces.

TACs often are responsible for aircraft deconfliction. When strike aircraft arrive on the scene, they check in with the TAC. Depending on the number of aircraft on station, the TAC may direct some aircraft to orbit nearby, assigning altitudes and locations to prevent midair collisions. The TAC assigns targets based on the aircrafts' capabilities (e.g., on-board sensors, endurance, weapons load) and their fuel states.

TAC-Aircraft Communications

When aircraft arrive on station, they use the "CAS check-in briefing" to notify the TAC. The briefing gives the mission number, number and type of aircraft, position and altitude, ordnance, time on station, and abort code. The TAC may put the aircraft into a holding orbit or give the nine-line briefing, which tells the pilot what type of control procedures are being used, then gives the (1) initial point (IP), (2) the heading/offset, (3) the distance to target from the IP in nautical miles, (4) the target elevation (in feet above mean sea level), (5) the target description, (6) the target location (in latitude/longitude, coordinates, or offsets or by a visual description), (7) the type of marking

---

port and/or Naval Air" (http://www.mfr.usmc.mil.hq/3danglico/3d%20ANGLICO.htm (last accessed November 7, 2003)).

used (white phosphorus, laser, infrared) and the laser code if applicable, (8) the location of friendly forces (from the target in meters and with a cardinal direction and an indication of how the position is marked), and (9) the egress route.[8]

Until recently, TACs often had to provide considerably more information to get aircraft on target. The TAC would help the fighter pilot find the target by identifying major terrain features such as rivers and towns. Beginning with these larger features, the TAC talk-on would reference increasingly small details until the target was in sight. Recognition panels, smoke, mirrors, and other signaling devices would aid in identifying friendly and enemy positions. Airborne TACs, flying small, slow aircraft such as the O-1 "Bird-Dog" used in Vietnam, fire small white-phosphorus rockets into the enemy positions, producing white smoke that is visible to the fighter pilot. The following is a Vietnam-era exchange that gives some sense of how involved the TAC talk-on could be:

> "Sabre 21 is a flight of two F-100s, mission number 2311. We're carrying eight 117 slicks, point oh-two-five, and 1,600 rounds of 20 mike-mike."
>
> "Rog, Sabre. I copy. Ready for target info?"
>
> "Rog."
>
> "OK, your target is a known VC location. We got some mortar fire out of here last night. Also, there is at least one .50 cal in the vicinity. I'm not being shot at now, but the FAC up here this morning took a hit. So you can expect auto weapons fire. Copy?"
>
> "Rog. Sabre 21 copies."
>
> "OK, the friendlies aren't too close to this target. There is a fire support base about 700 meters southwest of the target. When you get below the clouds, you'll be able to see it on a bald hilltop. Target elevation is 2,700 feet. We've got a pretty stiff wind

[8] Joint Chiefs of Staff, 2003, pp. V-22–V-24.

from the east, about 15 knots on the surface and at 2,000, and 20 knots at 5,000 and 7,000. Copy?"

"Rog. Do you have a preferred run-in heading?"

"Rog, Sabre. I don't want you to overfly friendlies. Make your runs from southeast to northwest, breaking right after your drop. That way, bomb smoke won't obscure the target on your run-in heading. Over."

"Rog, Cider 45 . . . I understand—friendlies 700 meters to our left as we attack from southeast to northwest . . . break right. I'm down below the clouds at the rendezvous point. Don't have you in sight. I think I have the fire support base in sight. Over."

"Rog, Sabre. I'm about one k north of that and I see you. I'm at your three o'clock low. I'm rocking my wings. Over."

"This is Sabre 21. Have you in sight. We're ready to go to work."

"Stand by, Sabre. I'm getting final clearance from the Army on FM."

"Standing by."

"OK, Sabre. We're ready now. If you have me in sight, the target is just off my right wing. Call me when you want a mark."

"Rog, Cider. I'm turning base now . . . go ahead. Sabre Flight, set 'em up . . . hot-arm, nose, and tail."

"Sabre 21, my smoke rocket is away. I'll hold to the south. I have you in sight. Do you see my smoke?"

"Rog. I have your mark. Am I cleared in wet?"

"You are cleared in wet. Hit ten meters to the right of my mark."

"Understand cleared in wet. I have you in sight. You want me to hit ten meters northeast of your mark. Two away. Sabre 21 is off right."

"Good hit, lead. Two, do you have lead's bomb?"

"Rog. I see it."

"OK, move yours up the hill 20 meters."

"Rog. Understand 20 meters at 12 o'clock."

"That's right. I have you in sight, on base. You are cleared in wet."

"Rog, Cider. I understand cleared in wet. I have you in sight to my left. . . . Two away—off right."

"OK. Good hit . . . outstanding. Now, Sabre flight, hold high and dry while I take a look."

"This is Sabre 21 . . . high and dry."[9]

The need for this type of talk-on has been greatly reduced with the advent of coded laser designators and pointers, aircraft sensor pods that can search for and lock onto specific laser codes or slew to GPS coordinates, and systems such as the Mark VII range-finder that combine a laser range-finder with compass and GPS to provide the coordinates of the enemy position and GPS-guided munitions.

The following exchange, related in an email[10] from an A-10 pilot flying out of Bagram Airbase in Afghanistan, took place one night between an A-10 aircraft two-ship formation (Misty 21) and a ground TAC (Fortune 18). It is typical of the current state of the art:

"Misty 21, this is Fortune 18 how copy?"

"Misty has you 3 by 5" [5 by 5 would be loud and clear]

"Misty 21 is two A-10s, mission number XXXX, carrying two by two Mark 82s, six rockets, one G model Maverick and a full gun of HEI."

"Fortune 18 copies. Misty, do you have my position?"

---

[9] Lester, 1997, pp. 135–139.

[10] By the time the email reached the authors of this report, its author's name had been stripped out. The description of tactics is consistent with those reported in official Air Force after-action reports.

"We think so. Rope north." [Misty is asking the TAC to direct his IR pointer in the direction of the aircraft and make a circle.]

"Misty 21 visual."

"Friendlies are my position, south and east 500 meters."

"Contact the friendlies." [Misty has visual observation of friendly forces.]

"Fortune was taking mortar and rocket fire from our northwest to southwest from up on a ridgeline. Stand by for sparkle."

"Fortune 18 requests ordnance on my sparkle."

"Fortune, Misty 21. What type of ordnance do you request?"

"Mark 82s. Your restricted run-in is from southeast to northwest. Additionally, you must not hit the backside of the hill. There are Afghani military forces in position there and they are unmarked."

"Misty 21 can be in in 30 seconds."

"Roger, Misty call in with direction, wings level on final."

"Sparkle on, spot, Misty is in hot."

"Misty is cleared hot."

"Misty will be in in five seconds."

"Perfect bombs two." [Ground TAC reports that the bombs hit the target.]

TAC Proficiency Standards and Training Requirements

It is clear from the preceding discussion that the TAC and forward air controller–airborne (FAC-A) roles are highly demanding. Air Force, Marine, Navy, and Army FAC-As are already highly qualified aircrew (fighter or helicopter pilots or weapons-systems officers) before they specialize in this role. Ground TACs in the Air Force are either ALOs, ETACs, or CCTs. ALOs are rated officers, often with airborne FAC experience. ETACs hold the Air Command and Control Specialist Air Force Specialty Code (AFSC) 1C4. It typically takes an ETAC two to four years to become a certified TAC. CCTs go through a year course that includes qualification as FAA-certified air traffic controllers. Ground TACs are required to do 12 "controls" of aircraft dropping live munitions per year. This is a minimum requirement; it takes more like 20 controls annually to be proficient.[11] Ground TACs struggle to get access to the fighter sorties and ranges to meet these requirements. Only A-10 and AC-130 units make this a priority, and AC-130 controls do not count toward the requirement. Often the training is stilted, with controllers taking turns doing canned controls from a tower. Rarely are live controls integrated into joint exercises with the Army.

As the services explore the possibility of training Army artillery forward observers or other personnel to be ground TACs, these training constraints need to be addressed. It may be difficult to generate sufficient training sorties to support a significantly larger ground TAC pool.

One alternative training method that might replace some of the live controls is the use of simulator facilities. Even though the simulator cannot replace live controls, it offers the potential for more-complex and realistic training scenarios. The simulator would be similar to training facilities used by civilian special weapons and tactics teams and military SOF. It would be essentially a large room with projector screens, laser sensors, and wireless computer interfaces on the walls. Terrain, structures, and friendly and enemy forces would all be projected on the walls. The ground TAC would use his usual equipment, modified to work in this environment. He would use laser pointers and designators, laser range-finder/GPS systems, radios, and other equipment to call for fires as he moves with the friendly ground force. He might also have to use his personal weapon, as ground TACs have done during recent conflicts. The strike aircraft could be simulated by a computer program, or the simulator could be

---

[11] This is the view of combat-experienced controllers we interviewed at the Commander's Conference, 18th Air Support Operations Group, at Pope Air Force Base, North Carolina, on July 23–25, 2003.

networked with aircraft simulators.[12] A few simulator facilities like this could provide varied and realistic training opportunities for controllers from all the services. Initial locations could include Pope Air Force Base, North Carolina (home of the 18th Air Support Operations Group, the 21st Special Tactics Squadron, and the 24th Special Tactics Squadron), Hurlburt Field, Florida (home of the 23rd Special Tactics Squadron), McChord Air Force Base, Washington (supporting the 1st Air Support Operations Group), and Nellis Air Force Base, Nevada (home of the Air-Ground Operations School).

The TAC Manning Dilemma

The demand for TACs is growing, for several reasons. In Operation Enduring Freedom, the United States achieved a stunning victory through the innovative integration of indigenous land forces, SOF, TACs, and modern air power. However, TACs were in short supply, with just one CCT assigned to each Operational Detachment Alpha (ODA). Second, the war on terrorism may require U.S. or allied land forces to conduct counterinsurgency operations (such as those ongoing in Afghanistan and Iraq) over a prolonged period. These operations typically require relatively small units (squads and platoons) to operate independently in areas where organic firepower may not be available to support. To ensure that air power can be effectively directed in support of these units, TACs may need to be assigned down to platoon level. In the not-too-distant future, TACs might even be required at squad level, as they currently are for Special Forces when the 12-man ODA operates independently. Finally, Army Transfor-

---

[12] A recent experiment between the 19th Special Operations Squadron (AC-130 Gunship School) at Hurlburt Field, Florida, and the Army Dismounted Battlelab at Fort Benning, Georgia, linked a Ranger element and an ETAC in the lab in Georgia with pilots and crew in AC-130 simulators in Florida. The simulation presented them all with the same terrain (the Hurlburt airfield). The Ranger mission was to parachute onto an airfield and seize it. As the enemy responded, the ETAC called in and adjusted fire from the AC-130s (Simulation-facility overview briefing by Lt. Col. Michael Plehn, Commander, 19th Special Operations Squadron, Hurlburt Field, Florida, July 29, 2003).

mation envisions Units of Action operating in a much more dispersed fashion than current heavy forces do.

Both counterterror and counterinsurgency operations will likely require U.S. air forces to work increasingly with indigenous ground forces. In the most likely venues for these operations (e.g., ungoverned territories or states facing insurgencies), the local friendly ground forces will not have their own TACs. Thus, these types of operations will increase the demand for TACs both within the U.S. military and from our allies.

Support for Army Special Forces

Over the past decade, Air Combat Command has provided ETACs and ALOs to the five U.S. Army Special Forces groups and the Ranger Regiment. These Air Force personnel were originally assigned to Special Forces groups to train Special Forces personnel to conduct emergency CAS, but they quickly took on operational responsibilities as well. With only 16 controllers or fewer assigned to each Special Forces group, two-man elements could at most support eight of the 54 ODAs in a group.[13] During Operation Enduring Freedom, the ODAs often conducted split operations, i.e., operated in two six-man sections. Each section needed a controller, but the TACs found they were most effective operating as two-man teams. In some cases, the TACs kept the two-man structure, but they then had to rotate between the sections.

Although no formal request has been made, the Army has suggested that two controllers should support each ODA. If every Special Forces group were supported at this level, the Air Force would have to provide 108 controllers per group, for a total of 540 TACs—460 more than are currently available.

---

[13] The ODA (or "A Team") is the basic operational element of a Special Forces Group. Each ODA is commanded by a captain, with a warrant officer usually as the second in command. The ten remaining members of the team are experienced noncommissioned officers with expertise in light and heavy weapons, explosives, communications, and medical care. The ODA team is designed to be split, if necessary, into two six-man elements.

TACs and the War on Terrorism

At this time, it is difficult to determine how much the war on terrorism will increase the requirement for controllers. Current operations in Afghanistan and Iraq require little CAS, so large numbers of TACs are not in the field simultaneously. But even so, these commitments are stressful, because the career fields are not manned to support indefinite deployments.

The more difficult question is whether the conflicts in these countries might evolve in ways that dramatically increase the need for CAS. If conflicts in these countries or elsewhere rose to a level of lethality equal to that of the Vietnam War, situations would ensue in which U.S. patrols might be ambushed and would risk annihilation unless strong covering fire were available. If, due to terrain and other factors, aircraft were the most effective means to provide this fire, the demand for TACs would increase dramatically.

Current Demand for TACs[14]

TACs are assigned to ASOSs associated with Army divisions, Special Forces groups, and the 75th Ranger Regiment. The total number of TACs required for an Army unit, which we count as two-man "TAC elements" because they operate in pairs (usually one certified TAC and one yet-to-be-certified ETAC), depends on the type of Army unit that is being supported. Heavy (armor and mechanized infantry) units receive fewer TACs than light (airborne and light infantry) units do.

While both types of units have TAC elements at each maneuver battalion headquarters, the light brigade is allocated twice as many TAC elements for its company-level units. Sufficient TAC elements are normally assigned to light brigades to allow two-thirds of the ma-

[14] The Air Force also has special tactics squadrons (STS) in the AFSOC. The STS have CCTs that conduct a variety of missions, including terminal attack control in support of special operations. This analysis is limited to considering how Army Transformation may affect the demand for conventional TACs. However, both TACs and CCTs draw from the same pool of resources to support initial qualification and currency training.

neuver companies to have a TAC element assigned. In heavy brigades, there are typically only enough TAC elements to accompany one-third of the maneuver companies.

The variability in allocations for different types of Army brigades is nominally based on the brigades' capabilities and style of operations, particularly their requirement for air support and the degree of dispersal during combat.[15] According to Air Force planning factors for each type of Army unit, the pre-Transformation (circa 2000) Army force structure required 292 two-person TAC elements.[16] Table 6.1 illustrates how the mix of brigades in the Army force structure combined with the number of TAC elements assigned to each type of brigade to generate this overall level of demand.

Future Demand for TAC Elements

Army Transformation will greatly change the way Army combat units are organized and the way they will operate. Organic fire support is likely to be more scarce than it was in past operations, which may imply greater reliance on air-delivered munitions even if precision-guided munitions fulfill their promise. Moreover, Army forces will operate in a far more dispersed manner, implying that they will often

Table 6.1
Requirement for TAC Elements in Pre-Transformation Army Structure

Type of Army Unit (brigades and brigade equivalents)	Number	TAC Elements Each	Total TAC Elements
Armored, mechanized infantry, cavalry brigade	18	6	108
Light infantry, mountain, airborne, air-assault brigade	15	9	135
Ranger regiment	1	9	9
Special Forces group	5	8	40
Total	39		292

NOTE: Number = number of such units in the current Army structure; TAC Elements Each = TACs aligned with each unit, counted as two-man teams.

[15] Interview with Air Combat Command planners, July 2003.

[16] Unit-type codes for air support operations squadrons.

be beyond range of artillery. Coupled with the reduced level of fire support available from Army channels, the dispersed scheme of maneuver will generate demand for substantially more TAC elements. Indeed, this trend has already been observed in recent operations featuring dispersed maneuver by conventional Army units, such as Operation Iraqi Freedom.[17] Additionally, Army planners have reportedly notified Air Force planners that, presumably because of trends in firepower and dispersion, the Army will likely request a TAC for each maneuver company in the Stryker brigades.[18]

Dispersal may imply that a TAC cannot provide support across a battalion front when he is located with one company of the battalion. Instead, he may be able to support only that one company. As a result, every company might need a TAC, generating a greatly increased demand. Although heavy brigades may operate in very fluid fashion, as they did during operation Iraqi Freedom, they traditionally mass combat power on fairly narrow fronts. Traditionally, a brigade would cover a 10- to 20-kilometer front, allowing two forward-deployed TAC teams to oversee much of the battle area. By contrast, a Stryker brigade is designed to operate over a much larger area, with its battalions widely dispersed, perhaps beyond range of mutual support. It might not have any discernible front, only areas where contact had been made or was anticipated. Each of its companies might be similarly dispersed, so that a TAC supporting one company would be out of touch with the remainder of the battalion.

The Army has not yet formally articulated its requirement for TACs with Stryker brigades, but it will probably be higher than the requirement for maneuver brigades in the current force. Within each Stryker brigade, the Air Force might be requested to provide a TAC team to the brigade headquarters, to the cavalry-squadron headquarters, to each battalion headquarters, and to each company-level infantry and reconnaissance element. On this assumption, each Stryker brigade would require 16 TAC elements. The total requirement for

---

[17] Interview with 15th ASOS planners, August 2003.

[18] Interview with Air Combat Command planners, July 2003.

all Army units would rise to 330 two-man teams when the Stryker brigades are fully formed in fiscal year 2007 (see Table 6.2).

As described in Chapter Four, the Army is also converting its non-Stryker brigades into BUAs and is creating at least ten additional BUAs. It therefore plans to field around 44 BUAs by the end of fiscal year 2006, split evenly between heavy and light BUAs.[19] Emerging doctrine indicates that the Unit of Action will operate in an even more dispersed manner than the Stryker brigade, possibly over hundreds of thousands of square kilometers.[20] Assuming that the Army will demand TAC elements with each maneuver-battalion headquarters and maneuver company in the new units, as with the Stryker brigade, the Air Force might be asked to provide 12 TAC elements for each brigade.[21] The total requirement for all Army units might

Table 6.2
Requirement for TAC Elements with Stryker Brigades

Type of Army Unit (brigades and brigade equivalents)	Number	TAC Elements Each	Total TAC Elements
Armored, mechanized infantry, cavalry brigade	17	6	102
Stryker brigade	5	16	80
Light infantry, mountain, airborne, air-assault brigade	11	9	99
Ranger regiment	1	9	9
Special Forces group	5	8	40
Total	39		330

NOTE: Number = number of such units in the current Army structure; TAC elements = TACs aligned with each unit, counted as two-man teams.

---

[19] Feikert, 2004, pp. 8–9.

[20] "Precision Fire and Maneuver," Briefing, U.S. Army Training and Doctrine Command Deputy Chief of Staff for Doctrine, Concepts, and Strategy, 2003, available at http://www.objectiveforce.army.mil.

[21] According to Army plans, the heavy BUA will have four mechanized infantry companies, four armor companies, three scout troops (company-sized reconnaissance units), and one surveillance troop. The light BUA will have six infantry companies, three troops of mounted scouts, one troop of dismounted scouts, and an intriguing combat support company that includes a variety of assets, including a mounted assault unit. See Feikert, 2004, pp. 8–9.

thereby double, to more than 650 two-man teams, by 2007 (see Table 6.3).

The planned conversions to Stryker brigades will add modestly to the requirement for TACs, but conversions to BUAs could add dramatically to the requirement, especially when heavy units begin to convert (see Figure 6.1).

Demand is likely to grow still further if, as expected, there is a requirement to place one TAC element with each Special Forces A-Team. Total demand with all these changes could near 900 teams (see Figure 6.2).

This projection is speculative because the Army has not formally decided how many TACs should be aligned with a Unit of Action, nor has it formally requested a TAC element for every Special Forces A-Team. Nevertheless, the Air Force clearly faces the very real possibility that TAC demand might skyrocket far beyond the service's ability to organize, train, and equip 1C4 personnel.

Table 6.3
Requirement for TAC Elements with Brigade Units of Action

Type of Army Unit (brigades and brigade equivalents)	Number	TAC Elements Each	Total TAC Elements
Heavy BUA	22	12	264
Light BUA	22	12	264
Stryker brigade	5	16	80
Ranger regiment	1	9	9
Special Forces group	5	8	40
Total	55		657

NOTE: Number = number of such units in the current Army structure; TAC elements = TACs aligned with each unit, counted as two-man teams.

Figure 6.1
Potential Demand for TACs to FY 2007

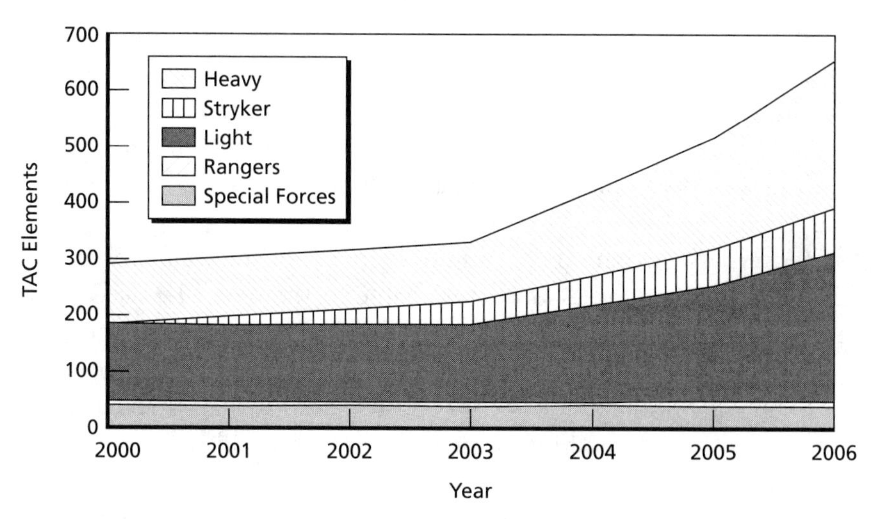

RAND *MG301-6.1*

Figure 6.2
Potential Demand for TACs to FY 2007, Including Potential Special Forces Requirement

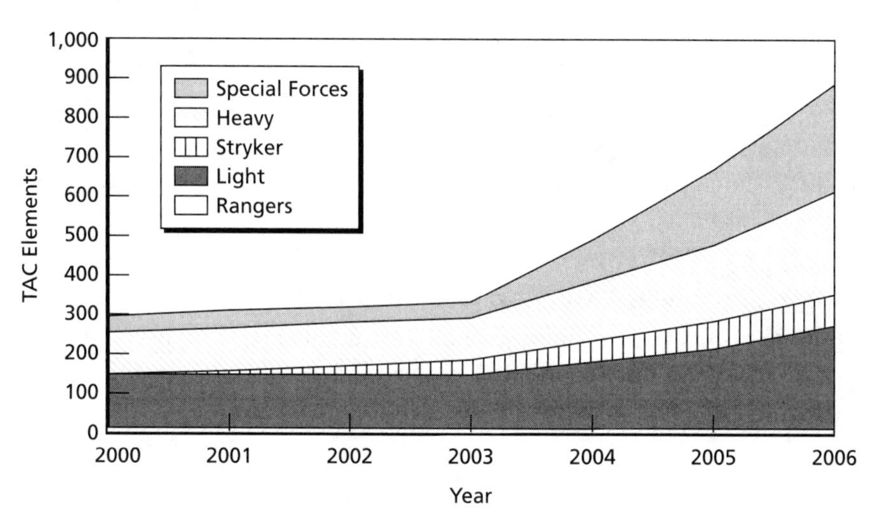

RAND *MG301-6.2*

New Concepts for Terminal Attack Control

Historically, the TAC preferred to have direct line of sight to friendly forces, enemy forces, and strike aircraft while controlling strikes. Having this vantage point allowed him to assure that strike aircraft would not endanger friendly forces and noncombatants. This deconfliction was most easily achieved when an airborne FAC was in charge.

Alternatively, the TAC might exercise indirect control. If the friendly or enemy forces were not visible to the TAC, he used radio to communicate with friendly ground forces, strike pilots, and the local ground tactical operations center (TOC). In situations requiring close support, indirect control was more tenuous than direct control and offered more possibilities for error, but it could be effective. Finally, ground units could employ emergency CAS procedures.

In this section, we explore new concepts and technologies that would allow TACs to control strikes effectively without having direct line of sight.

It is our judgment that these concepts are within the realm of the possible and merit further study. In particular, joint experiments and exercises, as well as additional research and development, will be necessary to determine if these are practical solutions and what mix of options is optimal. We also recognize that there are training, education, personnel, procedural, service culture, and possibly other paths and issues that need to be explored in the course of improving joint CAS.[22]

Expand Situational Awareness of Ground TACs

One possibility would be to give all TACs control of high-resolution, low-altitude, tactical UAVs.[23] This could expand the potential area that an individual TAC could be expected to cover (see Figure 6.3).

---

[22] We thank an anonymous U.S. Navy reviewer of the report for sharing this observation.

[23] AFSOC combat controllers used UAVs successfully in both Operation Enduring Freedom and Operation Iraqi Freedom (correspondence from the 720th Special Tactics Group, Hurlburt Field, Florida).

Figure 6.3
Expanding TAC Situational Awareness

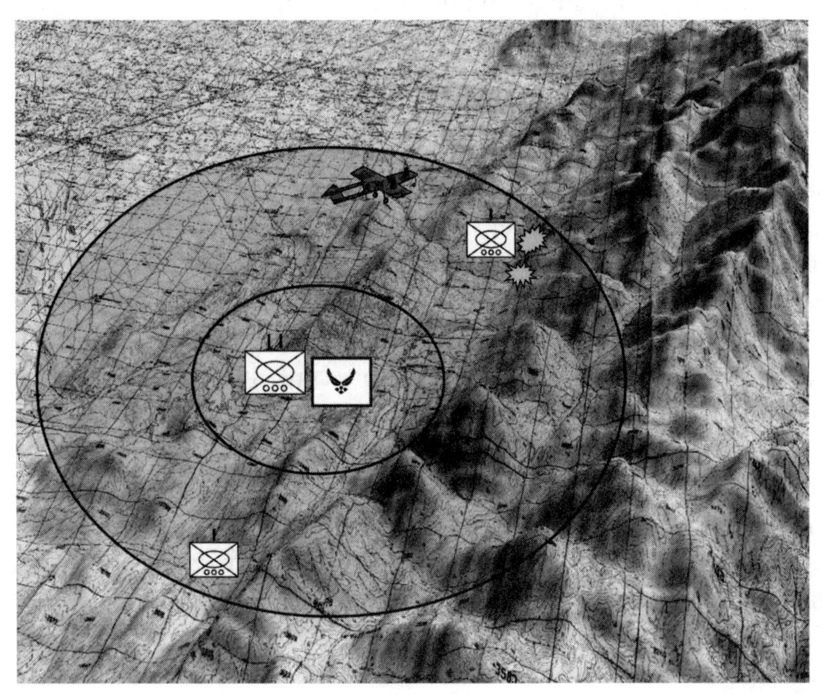

RAND *MG301-6.3*

TACPs could be equipped with UAVs or be able to task Army ground-launched or Air Force air-launched UAVs. When the Army's current plans come to fruition, UAVs will be routinely available at the tactical level. Depending on the tactical situation, the UAV might already be airborne to support an operation. More often, the UAV would be launched when friendly forces came into contact with enemy forces. The UAV would fly to the location of friendly forces. Once it arrived, the TAC would use the UAV's on-board sensors to understand the topography and tactical situation, beginning with the friendly forces and moving onto the enemy forces. The friendly land force might laser-designate or otherwise communicate the suspected locations of enemy forces. Once the TAC understood the tactical situation, he would call in aircraft and direct them to the target by providing coordinates, by lasing the target from the ground or from the UAV, or by conducting a traditional talk-on.

Using a UAV in this way would be challenging for the TAC; however, combat controllers are already using small UAVs successfully. The UAV sensors would likely provide either a high-resolution image with a narrow field or a low-resolution image with a wide field. The TAC would view the image on a small screen and would not benefit from the natural orientation afforded an airborne TAC or a ground TAC with direct sight of the target and friendly force. Indeed, viewing imagery collected from an aircraft moving at low altitude can be disorienting, and high resolution worsens this effect. To avoid this problem, the TAC might also receive still pictures of areas of interest.

A system designed for the TAC would integrate multiple sensor inputs, Blue Force Tracker data, and maps into a single display on a laptop computer. One option would be to integrate high- and low-resolution images on the screen in a way that mimics human sight. Human sight integrates a high-resolution image, where the eyes are focused, into a much wider low-resolution image, encompassing the area of peripheral vision. Similarly, the TAC's display would place high-resolution imagery within a lower-resolution image. Today, the display might employ separate windows. In the future, the display might integrate images from several cameras, seeking to provide a near-360-degree perspective.[24] The TAC might also receive a map display that showed the location of friendly forces, the location and flight path of the UAV, and locations of targets. To complete the picture, a link to the Airborne Warning and Control System (AWACS) or some kind of tracking system could provide data on strike aircraft, such as altitude, speed, approach path, weapons status, and remaining fuel.[25]

---

[24] This might require virtual-reality goggles or multiple screens. Such a system could work in a battalion tactical operations center (TOC). It remains to be seen whether it would be practical for dismounted TACs.

[25] Air Force experiments have already demonstrated a prototype system that allows a TAC to designate targets on a laptop map and send them digitally to strike aircraft. The laptop dis-

A networked battlefield should allow TACs to effectively control attacks from battalion headquarters and could perhaps reduce the requirement for company-level controllers.

Place TACs on Helicopters

Another option to expand a TAC's coverage would be to place him in an Army helicopter flying in armed reconnaissance (see Figure 6.4). For example, the 101st Airborne Division (Air Assault) allowed TACs

Figure 6.4
Putting TACs on Helicopters

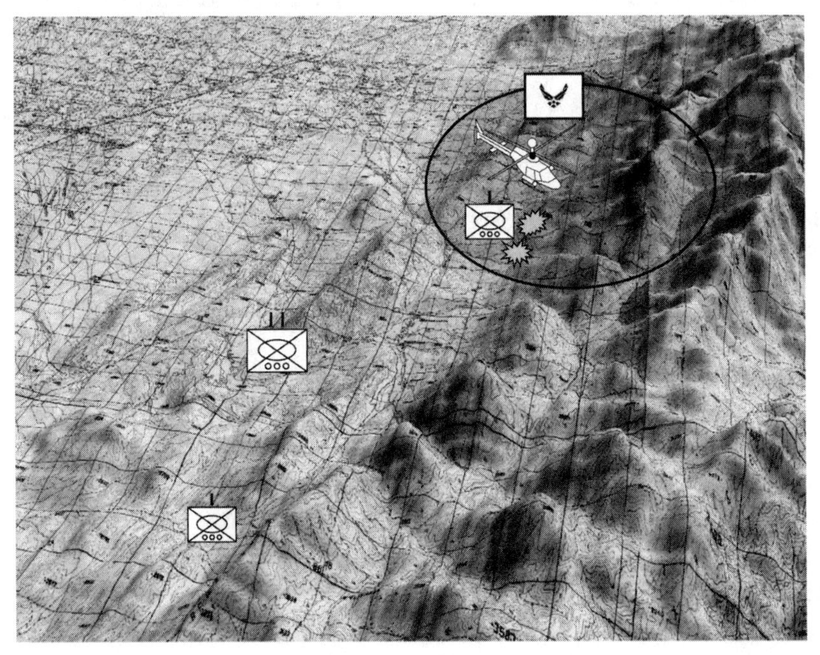

RAND *MG301-6.4*

---

play shows aircraft tracks, weapons loads, fuel state, and the targets the aircraft are engaging. In the experiment, when the TAC observed an aircraft reaching the IP (initial point) on his display, he would send a "cleared hot" signal digitally to the pilot (interview with Colonel T. C. Coon, USAF/XOX, February 24, 2004). Joint Forces Command has also recommended deploying the ALO Tactical Workstation (Joint Forces Command comments on a draft version of this report).

to fly in OH-58D Kiowa Warriors during Operation Iraqi Freedom.[26] This technique would give the TAC the perspective of an airborne FAC at low altitude, but he would have the additional advantage of being intimately acquainted with the terrain and the friendly commander's intention. In addition, there is something to be said for an airborne TAC who can land and discuss the situation face to face with friendly ground commanders. The TAC could fly either in a normal crew position or as an additional crew member. In a crew position, he would have access to communications and sensors, but at the cost of displacing a second pilot or weapons-systems operator. Alternatively, he might fly on a helicopter that provided room for a special communications and sensor package dedicated to his use.[27]

Use Helicopter Pilots as Airborne FACs

Another option would be to train helicopter crews to perform as FACs. A scout helicopter's agility, small size, and ability to fly low and slow and hover make it uniquely suited for airborne observation. Additionally, the most modern scout helicopters are equipped with excellent sensor suites. As a result, their crews can detect and identify some targets that would escape detection by fast-moving aircraft. Helicopter crews are already very busy, however, and they might have difficulty taking on an additional task, especially one as challenging as terminal attack control. They would need additional training, as discussed above. There are historic precedents for using helicopters as FAC-As. During the Vietnam War, scout helicopters performed FAC-like functions for attack helicopters and worked closely with Air Force FACs controlling fighter aircraft in joint air attacks.[28] More recently, Marine AH-1W Cobra pilots served as FAC-As during Operation Iraqi Freedom. Cobras laser-designated targets for fighter aircraft dropping laser-guided bombs. In one incident during Iraqi Freedom,

---

[26] Interview with Captain Paul "Dino" Murray, Air Liaison Officer, 101st Airborne Division, during the 18th Air Support Operations Group Commander's Conference at Pope Air Force Base, North Carolina, July 25, 2003.

[27] We thank Major Michael "Starbaby" Pietrucha, HQ USAF/XOXS, for sharing this idea.

[28] See Mills, 1992.

Cobras and a variety of fixed-wing aircraft attacked a bunker complex. One Marine section (four helicopters) hovered on line as they attacked the complex with Hellfire missiles, rockets, and 20-mm cannon. At the same time, the section leader was acting as a TAC, directing fighter aircraft unto the target.[29]

Clearly, Army and Marine scout and attack helicopter pilots already have many of the skills needed. We recommend that some percentage of them be given the additional training to become certified FAC-As. Although enemy air defenses will, at times, limit the use of helicopter FAC-As or necessitate less-than-ideal flight profiles, they offer an effective and even unique means to perform terminal control in many situations (see Figure 6.5).

Figure 6.5
Using Helicopter Pilots as FAC-As

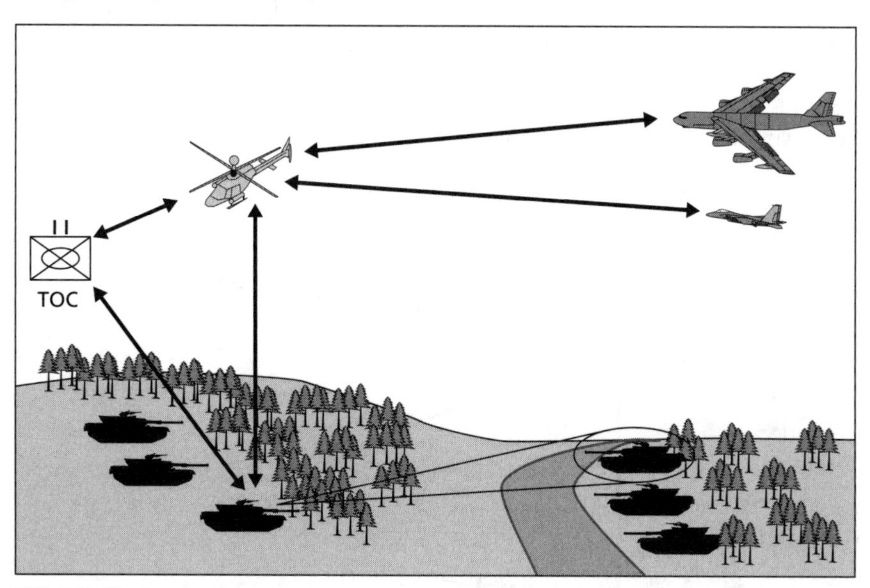

RAND *MG301-6.5*

[29] Maj. Jamie Cox, Operations Officer, Marine Light/Attack Helicopter Squadron 269, *A Personal Account of an AH-1W Pilot During the War with Iraq,* distributed by email within the Marine Corps and forwarded to RAND by U.S. Marine Corps Combat Development Center staff.

Enhance Capabilities of FAC-As

Another option would be to train more FAC-As and upgrade their equipment (see Figure 6.6). Due to the speed of their aircraft, FAC-As can cover huge amounts of territory. Rather than place ground TACs with many small units on a dispersed battlefield, the Air Force might supplement them with FAC-As. Currently, FAC-As primarily work at some distance from friendly ground forces, finding targets and directing strike aircraft onto them. In this option, some FAC-As would shift their focus to working closely with ground maneuver forces. They would communicate directly with soldiers, e.g., platoon sergeants and platoon leaders, who would need to have at least a basic understanding of aircraft and weapons. This concept would necessitate routine joint training between FAC-As and ground forces. In addition, FAC-As would need enhanced capabilities. For example, an enhanced OA-10 would have new sensors and digital links. A modern sensor pod, such as a Sniper or Litening, would allow FAC-As to

Figure 6.6
Enhancing the Capabilities of FAC-As

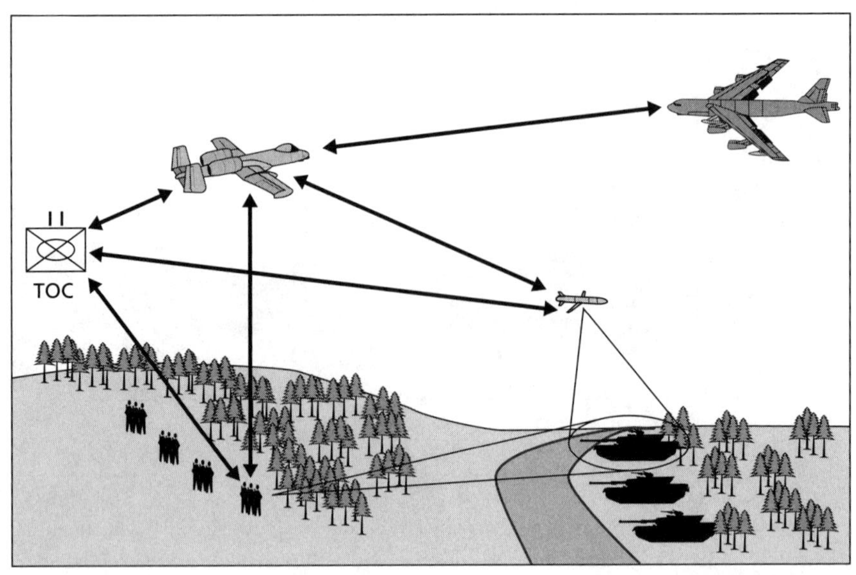

RAND *MG301-6.6*

automatically search for coded laser pointers, automatically slew to GPS coordinates, and search for targets day or night. Digital links to the Joint Surveillance and Target Attack Radar System (JSTARS), UAVs, strike aircraft, and land forces would give airborne TACs a multidimensional view of the battlefield. If all players were able to share imagery, targets could be identified quickly, reducing the risk of fratricide. Finally, an on-board mini-UAV would give the FAC-A an autonomous capability to take a very close, high-resolution look at a suspected target and would make it possible for a FAC-A flying above weather to see what is happening below the clouds.

Enhance Bombers as CAS Platforms

The bomber's long range, long endurance, and large payload made it ideal for operations in Afghanistan and extremely useful for operations in Iraq. Initial air operations over Afghanistan had to be conducted from remote land bases or from carriers in the Indian Ocean. As a result, naval aviation and bombers flew most of the strike missions. B-52 and B-1 bombers struck al Qaeda and Taliban forces throughout the country. In numerous battles, bombers conducted attacks that enabled Northern Alliance or other friendly land forces to rout enemy forces and capture key terrain. In some cases, bombers saved small friendly forces from being overrun by much larger Taliban forces.

What made the bomber so effective in Afghanistan was the combination of U.S. TACs on the ground, laser range-finders to determine the coordinates of enemy positions, and the GBU-31 JDAM, a GPS-guidance kit fitted onto the Mark 84 2000-lb bomb.[30] B-52 crews enthusiastically embraced this mission, and the bomber community is now placing more emphasis on CAS. The Air Force Reserve Command purchased Litening sensor pods for the B-52s in the 93rd Bomb Squadron to support this mission.

---

[30] Twelve GBU-31s are carried on wing pylons on the B-52 in addition to its internal bomb load. See http://www2.acc.af.mil/accnews/dec00/000401.html (last accessed November 13, 2003) and Air Armament Center, 2003, pp. 5-32–5-38.

Given the right equipment and training opportunities, bombers can make even greater contributions to future air-ground partnerships. For most of the Cold War, bomber crews were trained to fly to a set of coordinates and deliver their munitions against a fixed target at that location in all weather. Although they were heavily used during the Vietnam War to deliver conventional munitions, often fairly close to friendly forces, their primary mission and training focused on delivery of nuclear weapons against targets in the Soviet Union. Bomber crews rarely interacted with TACs. They were not typically a part of the decision to strike a particular target. They were given their target folders and expected to put weapons on the target accurately and at the specified time.

CAS requires a different mindset and more information than just target coordinates. Bomber crews must be trained to take part in an interactive process involving ground TACs, TAC-As, and reconnaissance aircraft such as JSTARS to learn where friendly forces are located, where enemy forces are suspected, and whether there are noncombatants in the area. They need to be more than passive recipients of targeting data.

The Air Force Reserve initiative to equip B-52s with Litening pods is an essential first step. The Litening pod gives the bomber crew a high-resolution electro-optical and infrared image of the target, enabling them to work with TACs using laser pointers and designators. With the Litening pod, a bomber crew can zoom in on coordinates, allowing them to examine a target before releasing precision munitions. Additional digital links would further enhance bomber-crew situational awareness. For example, a ground TAC could send the bomber crew an image from his perspective, and the bomber could send the ground TAC an image from its perspective. By comparing images, the TAC and the bomber crew could greatly speed and abbreviate the talk-on. The Marine Corps already has a downlink that allows Harrier aircraft equipped with the Litening pod to send an image to a ground TAC. The airborne TAC in an OA-10 benefits from access to JSTARS displays, Global Hawk imagery, and other information, and the B-52 crew should have access to similar feeds.

A more ambitious but still evolutionary approach would give some number of B-52s a gunship-like capability, implying the following characteristics:

- Digital links to TACs, ground forces, sensor platforms, and other strike aircraft
- A targeting pod, e.g., Sniper or Litening
- Air-droppable mini-UAVs
- Air-droppable unattended ground sensors
- A six-man crew
- A weapons suite

We have already discussed most of these enhancements. The six-man crew and the weapons suite are additional improvements.

A six-man crew would comprise the aircraft commander, co-pilot, electronic-warfare officer, offensive-weapons officer, off-board-sensor operator, and TAC. The off-board-sensor operator would use the vacant position formerly occupied by a gunner. He would launch, control, and monitor mini-UAVs and unattended ground sensors, which would allow the B-52 to see targets below a cloud ceiling and would produce high-resolution imagery. The aircraft would be equipped with radios and digital links capable of communicating with the engaged ground forces, and one of the crew members would be trained and responsible for communications with the ground commander.

A new weapons suite would take advantage of the B-52's ability to carry a large mix of munitions. For example, it might carry JDAMs externally while carrying the Small-Diameter Bomb and the Very Small Missile internally.[31] The Very Small Missile is a conceptual system proposed by RAND for the next-generation gunship. It would be

---

[31] Ideally, a new-design rotary launcher would enable the B-52 to carry a large variety of weapons; the crew would simply dial up whatever weapon they desired. In discussions with members of the 93rd Bomb Squadron during a visit to Barksdale Air Force Base, Louisiana, we were told that existing weapons racks could be used to carry several homogeneous stacks of weapon types internally, giving the crew a choice among two or three options at all times.

a 45-lb missile with a 12-lb warhead cruising at Mach 1.6. GPS guidance with an optional laser seeker would make the Very Small Missile effective against both stationary and moving targets. It would produce about the same effects as the 105-mm howitzer currently carried by the AC-130 aircraft, but from higher altitude and longer standoff range, while allowing engagement from all azimuths. Like rounds from the 105-mm howitzer, several Very Small Missiles could be in flight simultaneously.

Disaggregate the Terminal Attack Function[32]

This final concept is based on the observation that there is a strong trend in ground warfare toward dispersion and nonlinearity. If this is correct, ground forces are likely to increasingly operate fairly autonomously as companies and platoons. Such forces will need routine access to joint fires for support and will often team with air power to fight, fix, and defeat enemies who themselves have dispersed to avoid being detected and destroyed by overwhelmingly superior U.S. reconnaissance and strike assets.

Army Transformation initiatives embrace this vision of the future battlefield, and current plans envision TACs assigned to every maneuver company. Army interest in training their own TACs is further evidence of their determination to have better access to air power at lower echelons. The key question is whether a TAC in every company (or lower level) is necessary to achieve the seamless access to air power that the Army desires.

In our judgment, it will prove extremely difficult to create sufficient certified and fully proficient TACs to meet evolving Army needs. There is a real danger that TAC standards and battlefield effectiveness will degrade if a rapid increase in TAC numbers is mandated. Even if enough personnel with the necessary aptitude for this de-

---

[32] The discussion in this section applies to TACs in support of conventional operations. Special operations may require a different model. The most risky and important special operations are likely to continue to require dedicated TACs. In other operations, an airborne TAC model might be viable. The TAC-at-battalion-TOC model would likely need to be modified, given the unique command-and-control requirements and arrangements associated with SOF.

manding mission can be recruited, trained, and kept proficient to meet the one-per-company requirement—an unlikely prospect, in our view—the dispersion trend suggests that having TACs at the platoon level will soon be viewed as necessary. The desire of Army Special Forces to have two TAC elements per A-Team is consistent with this view.

What we envision instead is accepting these developments as inevitable and working to understand which tasks must be accomplished by a fully certified TAC and which could be assigned to other personnel. Is it really necessary for TACs to be assigned below the battalion level? Might a properly networked and equipped TAC working out of the battalion TOC partner with forward observers at lower echelons to deliver effective air power to engaged units? Table 6.4 lists major TAC responsibilities and how they might be reassigned in the future. This list is focused on terminal attack control and does not include higher-level functions such as the integration and synchronization activities of the brigade ALO.

How would this work in practice? A platoon requiring CAS would determine the target location with a laser range-finder/

Table 6.4
Disaggregating Terminal Attack Control Functions

Requires fully certified TAC	Deconflict aircraft in target area Assign aircraft to targets Select munitions Assign attack headings Clear aircraft to drop munitions
TAC and Battalion S-3	Deconflict air and land operations
Near-term concept: TAC, pilot, and forward observer	Identify targets Adjust aimpoints Assess battle damage
Far-term concept: Engaged combat element	Call for fire Identify and geolocate targets Adjust aimpoints Assess battle damage

imaging/GPS device (a user-friendly version of the current Mark VII or Viper devices).[33] He would use a handheld or laptop-based digital link to send a simple call for fire (location of target in latitude and longitude, altitude of target, description of target, type of marking used, if any, location of friendlies) and an image (if available) to the battalion TOC.

The battalion fire support element and TAC would validate the target, using their understanding of the tactical situation, current rules of engagement, and commander priorities. Laptops that display JSTARS and AWACS tracks, imagery, and Blue Force Tracker information give the TAC in the battalion TOC greatly enhanced situational awareness. Once the FSO and the TAC determine that no friendlies or noncombatants appear to be at the target location, the request for fire would be digitally transmitted to the appropriate source. In the case of aircraft, the TAC would modify the simple request for fire received from the engaged unit with additional information needed by the combat aircrew. This would include the target IP (the beginning point for attack runs), heading/offset, distance to target from the IP in nautical miles, attack heading, egress route, and a description of air-defense threats), as appropriate. The TAC would monitor aircraft position, speed, heading, fuel state, and weapons load via a Link 16 display on his laptop and would clear the aircraft hot as they cross the IP.[34] Voice communications could supplement digital links as necessary. Aircraft control functions would remain at all times the sole responsibility of a fully certified, proficient Air Force TAC located at the battalion TOC.

---

[33] Ideally, this device would be sufficiently user-friendly that platoon leaders (not just forward observers) would be able to make basic calls for fire. Certainly, technology is moving in this direction, as are joint concepts for the networked battlefield. For example, joint doctrine recognizes that in Type 2 and Type 3 CAS, the observer may be a scout, a combat observation and lasing team, a fire support team, a UAV, a SOF, or another asset with real-time information on targets. See Chairman, Joint Chiefs of Staff, 2003c, p. V-15.

[34] The Air Force has already demonstrated a laptop system that allows a TAC to digitally transmit a call for fire to a fighter four-ship element. The laptop displays aircraft speed, heading, altitude, weapons load, targets acquired, and fuel state (via Link 16). In the demonstration, the TAC was able to reassign targets and clear the aircraft hot, all via his laptop.

The combat aircrew would arrive on the scene with basic information about the location of friendly and enemy forces and air-defense threats displayed on their combat data-link display. This information, broadcast as part of a common operating picture, would not display the exact location of moving forces because of the delays in updating and synchronizing the thousands of computers in the network. To ensure that friendly and enemy forces have not intermingled or switched positions, strike aircraft would use the following approaches. First, they would switch to a battlefield view that displays the location of friendlies based on the radio signals the aircraft's own antennas are picking up from the nearest friendly units. These signals come straight from the ground force's Blue Force Tracker transmitters at the speed of light, thus avoiding the latency problem. Second, the aircraft could use its radar to interrogate radar tags on friendly vehicles and potentially on every soldier.[35] Finally, the strike aircraft could use their on-board optical sensors when possible to further verify that the target is hostile. The Blue Force Tracker information and/or radar tag interrogator are essential to the disaggregated TAC concept. In our judgment, the teaming of the engaged element, TAC, and FSE at the battalion TOC and aircrews equipped with significantly enhanced information would allow the TAC function to be disaggregated without losing battlefield effectiveness. Indeed, if these information sources can be integrated on multipurpose displays, there is every reason to expect that TAC effectiveness could improve on the future battlefield.

Finally, air-ground operations are likely to increasingly involve the use of ground fires to suppress air defenses or joint air-attack teams of fixed- and rotary-wing aircraft.[36] Locating the TAC in the battalion command post would enhance the integration of these joint

---

[35] This radar tag system has been developed by Sandia National Laboratories for the Army. The tag "is a normally passive (nontransmitting) device mounted on tanks and other ground vehicles. When a fighter or bomber radar 'paints' a tagged vehicle, the sensor attaches a bit of data to radar pulses being reflected back to their source. A distinctive icon presented on the aircraft's radar display tells a pilot where friendly forces are located" ("Radar Tag Designed to Reduce Battlefield Fratricide," 2004, p. 18).

[36] Thanks to Jeffrey McCausland for this observation.

assets. This model would, of course, require that the TAC and battalion FSE train together routinely.

Conclusion

TACs, whether airborne or on the ground, play a critical role in the effective employment of air assets in close air attack. The changing security environment, enemy responses to U.S. air power, and Army Transformation initiatives all are likely to increase the demands for TACs—possibly substantially—over the next decade.

Some expansion of the professional TAC cadre is possible and will probably be necessary. Yet, even assuming that Army personnel are trained as TACs, significant challenges must be overcome to expand the TAC force. The TAC function is demanding, and relatively few military personnel are capable of handling it. It takes years to become certified, and considerable training, including live controls of fighter aircraft, to remain proficient. Given other demands on aircraft, it is unlikely that the number of sorties available for TAC training can be dramatically increased in the near term. These constraints, which the Air Force has faced for years, will also limit Army efforts to create "universal controllers."

For these reasons, other options, such as those presented in this chapter, will be necessary to meet the demand for TACs.

Looking to the future, it is likely that technology will make it easier for ground forces to gain rapid access to air power. As the battlefield becomes more digitized, the reluctance of airmen to attack targets nominated by engaged ground forces will likely lessen. The proliferation of Blue Force Tracker technologies will give pilots greater assurance that they aren't attacking friendlies, and new, more user-friendly systems will allow platoon leaders to determine enemy coordinates and send them digitally to combat aircraft. As digital formats replace voice communication, requests from ground forces should more readily meet the needs of combat aircrews. In the extreme, network-centric warfare might dramatically reduce the need for TACs. More realistically, TACs will probably always play an important role, but new technologies could greatly enhance their ability to cover the battlefield.

CHAPTER SEVEN

Conclusions

In this report, we have presented an overview and analysis of the major challenges facing the Air Force and the Army as they work to better integrate air power and land power. This analysis is intended to help inform Air Force and Army efforts to address the following three key policy questions:

1. How should air attack and ground maneuver be integrated on the future battlefield?
2. How should the CAS terminal attack control function be executed?
3. How should ground maneuver/fires and air attack be deconflicted?

In this final section, we present our key findings and recommendations.

Key Findings

- **Army Transformation is increasing Army interest in air-delivered fire support.** As the Army seeks to become more strategically deployable and agile on the battlefield, it is reducing the weight of fires available to maneuver units. Although not yet fully detailed, the number of independent artillery brigades will shrink as the Army shifts manpower in those units to military police and other undermanned functions. In addition, operations are expected to increasingly center on independent brigades that will operate without or with less division and corps fire support. These factors, combined with a newfound Army confidence in the accuracy and responsiveness of air-delivered fires, will result in increased Army requests for CAS and interdiction missions.

- **Army Transformation will increase the demand for TACs.** Current joint procedures require that a certified TAC control aircraft conducting nonemergency CAS missions. For this reason, the Army is now planning on the assignment of TACs to every maneuver company, an action that would require a substantial increase in the number of certified TACs.
- **The joint terminal attack controller (JTAC) program is not designed to generate a large number of certified TACs.** The JTAC program was created to ensure that TAC standards are uniform across the services, not to produce a vast new pool of TACs. Whether TACs are trained at a joint school or produced by the services, the fundamental constraints remain the same: a shortage of qualified candidates, a demanding job that takes years to master, a shortage of training facilities (ranges and simulators), and heavy demands on strike aircraft that make it difficult for them to generate the necessary training sorties for more than the current TAC force.
- **Operational/technological trends and manpower realities, not service preferences, are at the heart of the TAC debate.** Some view the TAC debate as the latest event in a long struggle between airmen and soldiers over the control of air power. In our judgment, however, the debate is driven by operational and manpower realities, not service preferences. The Army recognizes a strong trend toward dispersion and is appropriately adapting its forces to operate in smaller elements dispersed across a larger battlefield. Such forces will need more ready and routine access to air power. The Air Force is correct in insisting that only fully certified, experienced, and proficient TACs have the authority to control aircraft.

- **Creative use of available technologies can free TACs to focus on essential functions and give engaged ground elements greater access to joint fires.** The Army doesn't really need TACs with every engaged combat unit. What it needs is a system that allows engaged elements to designate targets, TACs, and FSOs at the battalion level to confirm that no friendly forces are at the target locations, and aircrews to independently confirm that the targets are good. The technologies discussed in Chapter Six would allow such a system. These technologies already exist or are well along in development.
- **Disaggregating the TAC function is essential to ensuring that both Army and Air Force battlefield needs are met.** Identifying TAC functions that could be delegated to engaged combat units (e.g., target identification and geolocation) would ensure that dispersed ground elements can easily call for air support and would allow TACs to focus on those functions that require a fully certified controller (e.g., aircraft control and deconfliction). It is the only option that has a high probability of meeting Army needs without undue risk to ground and air forces.
- **Army organic fires remain the most efficient means to meet routine unplanned requests for fire support.** Giving engaged ground elements the ability to effectively call for precision fires against enemy forces is a necessary but not sufficient condition for responsive fire support. The fires must also be rapidly available. Army standards for fire support responsiveness are very high, with counterbattery fire expected to be delivered within three minutes and more-general fire support in five to ten minutes. This level of responsiveness is possible from the air for selected high-priority missions (e.g., the leading elements in a major offensive such as the 3rd Squadron, 7th Cavalry during Operation Iraqi Freedom or Special Forces direct-action missions), but it requires a huge force structure to sustain for prolonged operations over a large battle space. New concepts for long-range joint fires might meet some of these needs, but the most responsive systems (missiles) tend to be extremely costly and often in-appropriate for small-unit fire support needs; and

even these cannot meet single-digit response times unless they are relatively close to or have hypersonic speed. Therefore, the Army should retain sufficient organic fires to meet the routine fire support needs of dispersed units. Air forces are best used to directly attack enemy maneuver forces throughout the depth of the battlefield, to support selected forces at high risk, to partner with ground forces in planned offensive operations, and to act as a theater reserve.

- **Air attack and ground maneuver should be planned as mutually enabling activities.** "Close air support" is a poor term that implies a one-sided relationship. In modern combat, air and ground forces increasingly are operating in mutually enabling ways. This partnership should be encouraged. "Close air attack" is a more accurate description of what modern air forces do in partnership with ground elements. Whenever possible, air elements should be free to conduct deep operations against enemy maneuver forces, thereby isolating the battlefield. These operations have the potential to deny the operational level of maneuver to enemy motorized forces, preventing them from conducting offensive operations at the brigade or higher level. On the isolated battlefield, friendly ground forces can operate in smaller, more-dispersed elements, finding and fixing enemy forces that increasingly will operate in small elements to minimize their signature. Air and ground forces will attack these forces cooperatively, with air aggressively seeking enemy forces beyond the immediate line of sight of engaged friendly forces and also providing direct support to friendly forces as needed. Finally, in this vision, ground forces should do those things that they are uniquely able to do: capture and hold territory, find and control WMD, and enforce peace.

Recommendations for the Air Force and the Army

As we look to the future, the opportunities for effective partnering of air and ground forces are likely to grow significantly. We recommend that the Army and the Air Force work together to develop new concepts and technologies to speed this process.[1] In particular, training, education, and doctrine will need to be adapted to more smoothly integrate air attack and ground maneuver; the TAC function will need to be disaggregated and new processes developed to effectively designate targets while ensuring that essential oversight remains with the TAC and combat aircrew; and improved fire support control mechanisms will be needed to exploit the benefits of the digital battlefield and to get maximum benefit from the ability of air power to roam over the battlefield.

As adversaries adapt and move away from massed motorized forces operating in the open to dispersed, smaller forces exploiting difficult terrain, a well practiced and developed air-ground partnership will increasingly become a necessity.

---

[1] USAEUR (U.S. Army Europe) and USAFE (U.S. Air Forces in Europe) are currently working on a joint concept to train and exercise NATO/USEUCOM (U.S. European Command) personnel in the use of air and space power in a joint context (EUCOM and USAFE reviewers of an earlier draft of this report).

Bibliography

Adams, Thomas K., "Radical Destabilizing Effects of New Technologies," *Parameters*, Autumn 1998, pp. 99–111.

Air Armament Center, *2003–2004 Weapons File*, Eglin Air Force Base, Florida, 2003.

Air Force Doctrine Center, *Strategic Attack,* Air Force Doctrine Document 2-1.2, Washington, DC: Headquarters, Department of the Air Force, May 20, 1998.

Air Force Doctrine Center, *Counterland,* Air Force Doctrine Document 2-1.3, Washington, DC: Department of the Air Force, August 27, 1999.

Air Force Doctrine Center, *Air Warfare*, Air Force Doctrine Document 2-1, January 22, 2000.

Alberts, David S., and Daniel S. Papp (eds.), *Information Age Anthology*, Washington, DC: Department of Defense, Vol. 1, 1997, Vol. 2, 2000, Vol. 3, 2001.

Alberts, David S., *Information Age Transformation: Getting to a 21st Century Military*, Washington, DC: National Defense University Press, 2002.

Applegate, Melissa, *Preparing for Asymmetry: As Seen Through the Lens of Joint Vision 2020*, Carlisle, PA: Army War College Strategic Studies Institute, September 2001.

Arkin, William M., "The Crisis in Kosovo," *Human Rights Watch,* New York, available at http://hrw.org/reports/2000/nato/Natbm200-01.htm.

"Army Must Overhaul Commo, Distribution Systems, Kern Says," *Aerospace Daily*, September 9, 2003.

Arquilla, John, and David Ronfeldt (eds.), *In Athena's Camp: Preparing for Conflict in the Information Age*, Santa Monica, CA: RAND Corporation, MR-880, 1997.

Arquilla, John, and David Ronfeldt, *Networks and Netwars: The Future of Terror, Crime, and Militancy*, Santa Monica, CA: RAND Corporation, MR-1382-OSD, 2001.

Atkinson, Rick, *Crusade, The Untold Story of the Persian Gulf War*, New York: Houghton Mifflin Company, 1993.

Banerjee, Neela, "Violence in Iraq Spreads; Six British Soldiers Are Killed," *New York Times*, June 25, 2003.

Barry, John, "Exclusive: Osama bin Laden and the Mystery of the Skull," *Newsweek*, July 8, 2002.

Barry, John, and Evan Thomas, "The Kosovo Cover-Up," *Newsweek*, May 15, 2000, pp. 23–26.

Bellamy, Christopher, *The Evolution of Modern Land Warfare: Theory and Practice*, London: Routledge, 1990.

Belote, Howard D., *Once in a Blue Moon: Airmen in Theater Command*, CADRE Paper No. 7, Air University Press, Maxwell Air Force Base, 2000.

Biddle, Stephen, *Afghanistan and the Future of Warfare: Implications for Army and Defense Policy*, Carlisle, PA: Strategic Studies Institute, U.S. Army War College, 2002.

Biddle, Stephen, "Afghanistan and the Future of Warfare," *Foreign Affairs*, Vol. 82, No. 2, March/April 2003, pp. 31–46.

Bowden, Mark, *Blackhawk Down: A Story of Modern War*, New York, Atlantic Monthly Press, 1999.

Bowman, Stephen L., et al. (eds.), *Motorized Experience of the 9th Infantry Division 1980–1989*, Washington, DC: Headquarters 9th Infantry Division (Motorized), Fort Lewis, WA, June 1, 1989.

Burger, Kim, "Future Combat Systems: The Global Battlefield," *Jane's Defence Weekly*, August 27, 2003.

Burns, John F., "As Baghdad Empties, Hussein Is Defiant," *New York Times*, March 19, 2003, p. 1.

Byman, Daniel L., *Air Power as a Coercive Instrument,* Santa Monica, CA: RAND Corporation, MR-1061-AF, 1999.

Byman, Daniel L., and Matthew C. Waxman, "Kosovo and the Great Air Power Debate," *International Security*, Vol. 24, No. 4, Spring 2000, pp. 5–38.

CENTAF, *Operation IRAQI FREEDOM—By the Numbers*, April 30, 2003.

Chairman, Joint Chiefs of Staff, *Doctrine for Joint Interdiction Operations,* Joint Publication 3-03, Washington, DC, April 10, 1997a.

Chairman, Joint Chiefs of Staff, *The Joint Doctrine Encyclopedia,* Washington, DC, July 16, 1997b.

Chairman, Joint Chiefs of Staff, *Doctrine for Joint Fire Support*, Joint Publication 3-09, Washington, DC, May 12, 1998.

Chairman, Joint Chiefs of Staff, *Joint Vision 2020*, Washington, DC: Chairman Joint Chiefs of Staff, 2001a.

Chairman, Joint Chiefs of Staff, *Unified Action Armed Forces (UNAAF)*, Joint Publication 0-2, Washington, DC, July 10, 2001b.

Chairman, Joint Chiefs of Staff, *Doctrine for Joint Operations*, Joint Publication 3-0, Washington, DC, September 10, 2001c.

Chairman, Joint Chiefs of Staff, *Command and Control for Joint Air Operations*, Joint Publication 3-30, Washington, DC, June 5, 2003a.

Chairman, Joint Chiefs of Staff, *Department of Defense Dictionary of Military and Associated Terms,* Joint Publication 1-02, Washington, DC, April 12, 2001, as amended through June 5, 2003b.

Chairman, Joint Chiefs of Staff, *Joint Tactics, Techniques, and Procedures for Close Air Support (CAS)*, Joint Publication 3-09.3, Washington, DC, September 3, 2003c.

Chandrasekaran, Rajiv, "Iraqi Mob Killed Britons," *Washington Post,* June 26, 2003.

Chandrasekaran, Rajiv, "Key General Criticizes April Attack in Fallujah," *Washington Post,* September 13, 2004, p. A17.

Cheek, Gary, "Why Can't Joe Get the Lead Out?" *Field Artillery Magazine,* January–February 2003, p. 33.

Clark, Wesley K., *Waging Modern War*, New York: Public Affairs (Perseus Books Group), 2001.

Clark, Wesley K., Gen. (USA), and Brig. Gen. (USAF) John Corley, "Press Conference on the Kosovo Strike Assessment," Brussels, Belgium: Headquarters, North Atlantic Treaty Organization, September 16, 1999.

Congressional Budget Office, *The Army's Bandwidth Bottleneck*, Washington, DC: Congressional Budget Office, 2003.

Conversino, Mark J., "The Changed Nature of Strategic Air Attack," *Parameters*, Winter 1997–98, pp. 28–41.

Cooling, Benjamin Franklin (ed.), *Case Studies in the Development of Close Air Support*, Washington, DC: Office of Air Force History, 1990.

Cordesman, Anthony, "Lessons of the Iraq War: Main Report," Washington, DC: Center for Strategic and International Studies, available at http://www.csis.org/features/iraq_instantlessons.pdf.

Corum, James S., and Wray R. Johnson, *Air Power in Small Wars*, Lawrence, KS: University Press of Kansas, 2003.

Daalder, Ivo H., and Michael E. O'Hanlon, *Winning Ugly, NATO's War to Save Kosovo*, Washington, DC: Brookings Institution Press, 2000.

Darilek, Richard et al., *Measures of Effectiveness for the Information Age Army*, Santa Monica, CA: RAND Corporation, MR-1155-A, 2001.

Davis, Lynn, and Jeremy Shapiro (eds.), *The Army and the New National Security Strategy*, Santa Monica, CA: RAND Corporation, MR-1657, 2003.

Department of the Air Force, *Operation Iraqi Freedom—By the Numbers*, U.S. Central Command Air Forces, Assessment and Analysis Division, Prince Sultan Air Base, Saudi Arabia, April 2003a.

Department of the Air Force, *Flying Operations, General Flight Rules*, Air Force Instruction 11-202, Vol. 3, Washington, DC, June 6, 2003b.

Department of Defense, "Executive Summary of the Battle of Takur Ghar," Washington, DC, May 24, 2002.

Department of Defense, *DoD Dictionary of Military and Associated Terms*, Washington, DC, 2003.

Deptula, David A., "Transforming Joint Air-Ground Operations for the 21st Century Battlespace," *Field Artillery*, July–August 2003, pp. 21–25.

Deptula, David A., Gary Crowder, and George Stamper, "Direct Attack: Enhancing Counterland Doctrine and Joint Air-Ground Operations," *Air and Space Power Journal*, Winter 2003, pp. 5–12.

Dinackus, David, *Order of Battle: Allied Ground Forces of Operation Desert Storm*, Central Point, OR: Hellgate Press, 2000.

Dixon, Robyn, "Kidnap Gangs Add to Iraqis' Insecurity," *Los Angeles Times*, August 6, 2003.

Dobbs, Michael, and Karl Vick, "Scores of Refugees Killed on the Road; NATO Says Jets Aimed at Military," *Washington Post*, April 15, 1999, p. A1.

Donnelly, John, "How US Strategy in Tora Bora Failed," *Boston Globe*, February 10, 2002.

Douhet, Giulio, *The Command of the Air*, Dino Ferrari (trans.), Washington, DC: Office of Air Force History, 1942, 1983 (reprint).

Echevarria, Antulio, and John Shaw, "The New Military Revolution: Post-Industrial Change," *Parameters*, Winter 1992–1993.

Escoto, Chantal, "Fourth Brigade to Add 2,000 Soldiers, Humvee," *Clarksville Leaf-Chronicle*, June 11, 2004.

Fattah, Hassan, "Random Death," *New Republic*, August 11, 2003, p. 14.

Feikert, Andrew, *U.S. Army's Modular Redesign: Issues for Congress*, Washington, DC: Congressional Research Service, July 19, 2004.

Filkins, Dexter, "Turkey Will Seek a Second Decision on a G.I. Presence," *New York Times*, March 3, 2003.

Filkins, Dexter, and John F. Burns, "Tentative Accord Reached in Najaf to Halt Fighting," *New York Times*, August 27, 2004, p.1.

Final Report to the Prosecutor by the Committee Established to Review the NATO Bombing Campaign Against the Federal Republic of Yugoslavia, Part III, The Hague, Netherlands, 2000.

Finney, Robert T., *History of the Air Corps Tactical School: 1920–1940*, Washington, DC: Center for Air Force History, 1992.

Forney, Matthew, "Inside the Tora Bora Caves," *Time*, December 11, 2001.

Freedman, George and Meredith, *The Future of War: Power, Technology, and American World Dominance in the 21st Century*, New York: Crown, 1997.

Fulghum, David A., "VTOL Transport Planned," *Aviation Week & Space Technology*, September 22, 2003.

Futrell, Robert, *The United States Air Force in Korea 1950–1953*, Office of Air Force History, Washington, DC, 1983.

Galloway, Joseph L., "Franks: We Held 25% of Iraq Before War," Interview with General Franks, *Miami Herald*, June 20, 2003.

General Motors General Dynamics Land Systems Defense Group, "Stryker Family of Vehicles," Fact Sheet, Washington, DC, November 2001. Available at: http://www.army.mil/features/stryker/stryker_spec.pdf.

Gompert, David C., *National Security in the Information Age*, Santa Monica, CA: RAND Corporation, RP-736, 1998.

Gordon, John, and David Orletsky, "Moving Rapidly to the Fight," in Lynn Davis and Jeremy Shapiro (eds.), *The U.S. Army and the New National Security Strategy*, Santa Monica, CA: RAND Corporation, 2003.

Gordon, Michael R., "The Goal Is Baghdad, But at What Cost, *New York Times*, March 25, 2003.

Graham, Bradley, "Bravery and Breakdowns in a Ridgetop Battle," *Washington Post*, May 24, 2002a, pp. A1, A16.

Graham, Bradley, "A Wintry Ordeal at 10,000 Feet," *Washington Post*, May 25, 2002b, pp. A1, A16.

Graham, Bradley, "Plans Made for Shift to Alternative Sites," *Washington Post*, March 3, 2003.

Grossman, Elaine M., "Jumper: Army, Air Force Work to Avoid Repeat of Anaconda Lapses," *Inside the Pentagon*, February 27, 2003, p. 1.

Hagenbeck, Franklin L., "Fire Support for Operation Anaconda," Interview by Robert H. McElroy, *Field Artillery*, September–October 2002, pp. 5–7.

Hallion, Richard, *Strike from the Sky: The History of Battlefield Air Attack 1911–1945*, Washington, DC: Smithsonian Institution Press, 1989.

Hannah, Mark, *Force XXI: The Army's Digital Experiment*, Strategic Forum Number 119, Washington, DC: National Defense University, July 1997.

Harrison, Marshall, *A Lonely Kind of War*, Novato, CA: Presidio Press, 1989.

Hawkins, Glen R., and James J. Carafano, *Prelude to Army XXI: U.S. Army Division Redesign Initiatives and Experiments, 1917–1995*, Washington, DC: U.S. Army Center for Military History, 1997.

Headquarters, Department of the Army, *Fire Support in the Airland Battle,* Field Manual 6-20, Washington, DC, 1988.

Headquarters, Department of the Army, *Tactics, Techniques, and Procedures for Fire Support for Corps and Division Operations,* Field Manual 6-20-30, Washington, DC, 1989.

Headquarters, Department of the Army, *Fire Support for Brigade Operations (Heavy),* Field Manual 6-20-40, Washington, DC, 1990a.

Headquarters, Department of the Army, *Fire Support for Brigade Operations (Light),* Field Manual 6-20-50, Washington, DC, 1990b.

Headquarters, Department of the Army, *Staff Officer's Field Manual Organizational, Technical, and Logistical Data Planning Factors (Volume 2)*, Field Manual 101-10-1/2, Washington, DC, 1990c.

Headquarters, Department of the Army, *Tactics, Techniques, and Procedures for Corps Artillery, Division Artillery, and Field Artillery Brigade Operations*, Field Manual 3-09.22, Washington, DC, January 1993.

Headquarters, Department of the Army, *Aviation Brigades*, Field Manual 1-111, Washington, DC, October 27, 1997.

Headquarters, Department of the Army, *The Army Vision: Soldiers on Point for the Nation, Persuasive in Peace, Invincible in War*, Washington, DC, 1999.

Headquarters, Department of the Army, *Air Cavalry and Troop Operations,* Field Manual 1-114, Washington, DC, 2000a.

Headquarters, Department of the Army, *The Infantry Brigade,* Field Manual 7-30, Washington, DC, 2000b.

Headquarters, Department of the Army, *Concepts for the Objective Force*, Chief of Staff White Paper, Washington, DC, 2001a.

Headquarters, Department of the Army, *Tactics, Techniques, and Procedures for the Field Artillery Battalion,* Field Manual 3-09.21, Washington, DC, 2001b.

Headquarters, Department of the Army, *Tactics, Techniques, and Procedures for Corps Artillery, Division Artillery, and Field Artillery Brigade Operations,* Field Manual 3-09.22, Washington, DC, 1993, reissued 2001c.

Headquarters, Department of the Army, *The Stryker Brigade Combat Team; Tactical Operational and Organizational Concept for the Maneuver Unit of Action*, TRADOC Pamphlet 525-3-90, Washington, DC, draft dated November 7, 2001d.

Headquarters, Department of the Army, *The Army Field Artillery Investment Strategy*, Washington, DC, 2002a.

Headquarters, Department of the Army, *Tactics, Techniques, and Procedures for Fire Support to the Combined Arms Commander,* Field Manual 3-09.22, Washington, DC, 2002b.

Headquarters, Department of the Army, *The Objective Force 2015*, Washington, DC, December 8, 2002c.

Headquarters, Department of the Army, *The Army Modernization Plan 2003,* Washington, DC, 2003a.

Headquarters, Department of the Army, *Army Posture Statement 2003*, Washington, DC, February 2003b.

Headquarters, Department of the Army, *Aviation Brigades,* Field Manual 3-04.111, Washington, DC, 2003c.

Headquarters, Department of the Army, *Field Artillery Investment Strategy*, G-8 Precision Strike Division, Washington, DC, 2003d.

Headquarters, Department of the Army, *The Stryker Brigade Combat Team*, FM 3-21.31, Washington, DC, 2003e.

Headquarters, Department of the Army, *United States Army Transformation Roadmap*, Washington, DC, 2003f.

Headquarters, Department of the Army, *Army Posture Statement 2004*, Washington, DC, 2004.

Headquarters, Department of the Navy, *Vision Presence Power*, Washington, DC, 2002.

Headquarters, 3rd Infantry Division (Mechanized), *Operation Iraqi Freedom, After Action Report*, Fort Stewart, GA, May 12, 2003.

Headquarters, United States Air Force, *Global Vigilance, Reach, & Power: America's Air Force Vision 2020*, Washington, DC, 2003.

Hersh, Seymour, "Offense and Defense," *New Yorker*, April 7, 2003.

Hornburg, Hal M., "Strategic Attack," *Joint Forces Quarterly*, Autumn 2002, pp. 62–67.

Hosmer, Stephen T., *The Conflict Over Kosovo: Why Milosevic Decided to Settle When He Did*, Santa Monica, CA: RAND Corporation, MR-1351-AF, 2001a.

Hosmer, Stephen T., *Operations Against Enemy Leaders*, Santa Monica, CA: RAND Corporation, MR-1385-AF, 2001b.

Howard, Michael, *Studies in War and Peace*, London: Temple Smith, 1970.

Hughes, Thomas A., *Overlord: General Pete Quesada and the Triumph of Tactical Air Power in World War II*, New York: The Free Press, 1995.

Hutcheson, Joshua, "The Transforming Army," *Army News Service*, April 22, 2004a.

Hutcheson, Joshua, "The Future of the 101st," *Army News Service*, April 29, 2004b.

Jane's All the World's Aircraft 2004–2005, Paul Jackson (ed.), Coulsdon, Surrey, UK: Jane's Information Group, 2004.

Johnson, Paul, *When Is Air Support to Special Operations Special?* Maxwell Air Force Base, AL: Air War College Research Paper, 2003.

Joint Chiefs of Staff, *Joint Tactics, Techniques and Procedures for Close Air Support (CAS)*, Joint Publication 3-09.3, Washington, DC, September 3, 2003.

Joint Staff, *Department of Defense Dictionary of Military and Associated Terms*, Joint Publication 1-02, Washington, DC, September 25, 2002.

Jones, Archer, *The Art of War in the Western World*, London: Harrap, 1987.

Kaufman, Gail, "USAF to Test Cargo UAV," *Defense News*, September 15, 2003.

Kraft, Nelson, "Lessons Learned from a Light Infantry Company During Operation Anaconda," *Infantry Magazine*, Summer 2002, p. 28.

Krepinevich, Andrew, "The Army and Land Warfare: Transforming the Legions," *Joint Forces Quarterly*, Autumn 2002, p. 76.

Lambeth, Benjamin S., *The Transformation of American Airpower*, Ithaca, NY: Cornell University Press, 2000.

Lambeth, Benjamin S., *NATO's Air War for Kosovo, A Strategic and Operational Assessment*, Santa Monica, CA: RAND Corporation, MR-1365-AF, 2001.

Lambeth, Benjamin, *Air Power Against Terror, America's Conduct of Operation Enduring Freedom*, Santa Monica, CA: RAND Corporation, MG-166-CENTAF (forthcoming).

Lester, Gary Robert, *Mosquitoes to Wolves: The Evolution of the Airborne Forward Air Controller*, Maxwell Air Force Base, AL: Air University Press, 1997.

Lewis, Michael, and Lt. Gen. Ned Almond, *USA: A Ground Commander's Conflicting View with Airmen over CAS Doctrine and Employment*, Maxwell Air Force Base, AL: Air University Press, 1997.

Mark, Edward, *Aerial Interdiction in Three Wars*, Washington, DC: Center for Air Force History, 1994.

Mayer, Jane, "The Search for Osama," *The New Yorker*, August 4, 2003.

Mazaar, Michael I., *The Military-Technical Revolution: A Structural Framework*, Washington, DC: Center for Strategic and International Studies, 1993.

McCaffrey, Terrance J., "What Happened to BAI? Army and Air Force Battlefield Doctrine Development from Pre-Desert Storm to 2001," unpublished thesis, School of Advanced Airpower Studies, Maxwell Air Force Base, AL: Air University, June 2002.

McDaniel, John R., *C2 Case Study: The FSCL in Desert Storm*, National Defense University, Command and Control Research Program, March 8, 2001, available at http://www.dodccrp.org/sm_workshop/pdf/C2_FSCL_doc.pdf.

McGrory, Daniel, and Michael Evans, "They Refused to Flee the Mob—Duty Made Them Stay Behind the Doorway to Death," *London Times*, June 26, 2003.

McKenzie, Kenneth, *The Revenge of the Melians: Asymmetric Threats and the Next QDR*, Washington, DC: National Defense University, McNair Paper 62, 2000.

McNeil, William H., *The Pursuit of Power: Technology, Armed Force, and Society Since A.D. 1000*, Chicago, IL: University of Chicago Press, 1982.

Meigs, Montgomery C., "Unorthodox Thoughts About Asymmetric Warfare," *Parameters*, Summer 2003, pp. 4–13.

Meilinger, Phillip S. (ed.), *The Paths of Heaven: The Evolution of Airpower Theory*, Maxwell Air Force Base, AL: Air University Press, 1997.

Metz, Steven, and Raymond Millen, *Future War / Future Battlespace: The Strategic Role of American Landpower*, Carlisle, PA: Army War College Strategic Studies Institute, 2003.

Mills, Hugh, Jr., *Low Level Hell: A Scout Pilot in the Big Red One*, Novato, CA: Presidio Press, 1992.

Ministry of Defence, *AP 3000: British Air Power Doctrine*, 3rd ed., London, 1999.

Moore, Harold G., and Joseph L. Galloway, *We Were Soldiers Once . . . And Young, Ia Drang: The Battle That Changed the War in Vietnam*, New York: Random House, 1992.

Moore, Robin, *The Hunt for Bin Laden*, Task Force Dagger, New York: Random House, 2003.

Moseley, Lt. Gen. (USAF) T. Michael, *Operation IRAQI FREEDOM—By the Numbers*, Shaw Air Force Base, SC: U.S. Central Command Air Forces, Assessment and Analysis Division, April 30, 2003.

Mueller, Karl P., "Strategies of Coercion: Denial, Punishment, and the Future of Air Power," *Security Studies*, Vol. 7, No. 3, Spring 1998, pp. 203–207.

Mueller, Karl, "Politics, Death, and Morality in US Foreign Policy," *Air and Space Power Journal*, September 2000, pp. 12–16.

Mueller, Karl P., "Threatening What the Enemy Values: Punitive Disarmament as a Coercive Strategy," in John Andreas Olsen (ed.), *Asymmetric Warfare,* Trondheim, Norway: Royal Norwegian Air Force Academy, 2002, pp. 117–142.

Nardulli, Bruce R., et al., *Disjointed War: Military Operations in Kosovo, 1999*, Santa Monica, CA: RAND Corporation, MR-1406-A, 2002.

Naylor, Sean D., "Air Force Policy Left Ground Troops High and Dry," *Army Times*, September 30, 2002, p. 10.

North Atlantic Treaty Organization, "Transcript of Press Conference by Mr. Jamie Shea and Brig. Gen. Giuseppe Marani," Brussels, Belgium: NATO, April 16, 1999.

Nygren, Kip P., "Emerging Technologies and Exponential Change: Implications for Army Transformation," *Parameters*, Summer 2002, pp. 86–99.

Office of the Secretary of Defense, *Transformation Planning Guidance*, Washington, DC, 2003.

Oliveri, Frank, "Heroes at Mogadishu," *Air Force Magazine*, June 1994 (accessed online on November 6, 2003 at http://www.afa.orga/magazine/June1994/0694gyros.asp).

Olsen, John Andreas (ed.), *Asymmetric Warfare*, Trondheim, Norway: Royal Norwegian Air Force Academy, 2002.

Olson, Mancur, Jr., "The Economics of Target Selection for the Combined Bomber Offensive," *RUSI Journal*, Vol. 107, November 1962, pp. 308–314.

Organization for Security and Cooperation in Europe, *Kosovo, As Seen, As Told*, December 1999.

Pape, Robert A., *Bombing to Win: Air Power and Coercion in War*, Ithaca, NY: Cornell University Press, 1996.

Peltz, Eric, et al., *Strategic Responsiveness: Rapid Deployment of Mission Tailored Capabilities for Prompt Power Projection,* Santa Monica, CA: RAND Corporation, AB-689-A, 2003.

Pengelley, Rupert, "Firing for Effect—Modern Artillery Strives to Deliver," *Jane's International Defense Review*, October 2003.

Peters, John E., et al., *European Contributions to Operation Allied Force, Implications for Transatlantic Cooperation*, Santa Monica, CA: RAND Corporation, MR-1391-AF, 2001.

Priest, Dana, "NATO Concedes Its Bombs Likely Killed Refugees," *Washington Post*, April 20, 1999, p. A19.

Priest, Dana, *The Mission*, New York, W.W. Norton & Company, 2003.

"Radar Tag Designed to Reduce Battlefield Fratricide," *Aviation Week and Space Technology*, March 15, 2004.

Record, Jeffrey, "Force-Protection Fetishism: Sources, Consequences, and (?) Solutions," *Air and Space Power Journal*, September 2000, pp. 4–11.

Record, Jeffrey, "Collapsed Countries, Casualty Dread, and the New American Way of War," *Parameters*, Carlisle, PA: Army War College, Summer 2002, pp. 10–12.

Riggs, Lt. Gen. (USA) John, "Transforming to the Objective Force," Briefing to the Atlanta Chapter of the Association of the U.S. Army, February 12, 2003.

Risen, James, "U.S. Officials Are Confident That Bunker Held Saddam Hussein," *New York Times*, March 22, 2003.

Romjue, John L., *From Active Defense to Airland Battle: The Development of Army Doctrine 1973–1982*, Fort Monroe, VA: U.S. Army Training and Doctrine Command, 1984.

Romjue, John L., *American Army Doctrine for the Post-Cold War*, Fort Monroe, VA: U.S. Army Training and Doctrine Command Historical Monograph Series, 1997a, pp. 2, 119–120.

Romjue, John L., *The Army of Excellence: The Development of the 1980s Army*, Fort Monroe, VA: U.S. Army Training and Doctrine Command, 1997b.

Rose, David, "Bloody Search for DNA to Discover bin Laden's Fate," *The Observer*, London, January 13, 2002.

Royal Australian Air Force Aerospace Centre, *AAP 1000: Fundamentals of Australian Aerospace Power*, 4th ed., Fairbairn, Australia, 2002.

Sanger, David E., "President Says Military Phase in Iraq Has Ended," *New York Times*, May 2, 2003.

Sanger, David E., and John F. Burns, "Bush Orders an Assault and Says Americans Will Disarm Foe," *New York Times*, March 20, 2003, p. 1.

Scales, Robert H., Jr., *Firepower in Limited War*, Novato, CA: Presidio Press, 1994.

Scales, Robert H., "Adaptive Enemies: Achieving Victory by Avoiding Defeat," in Robert H. Scales (ed.), *Future Warfare Anthology, Revised*

Edition, Carlisle, PA: Army War College Strategic Studies Institute, June 2001.

Scarborough, Rowan, "Hovering Spy Plane Helps Rout Iraqis," *Washington Times*, April 3, 2003.

Schwarzkopf, H. Norman, *It Doesn't Take a Hero*, New York: Bantam Books, 1992.

Sheftick, Gary, "3rd ID Testing New 'Unit of Action' at National Training Center," *Army News Service*, April 6, 2004.

Sheridan, Mary Beth, "Ground Fire Repels Copter Assault, *Washington Post*, March 25, 2003.

Silber, Laura, and Allan Little, *Yugoslavia, Death of a Nation*, New York: Penguin Books, 1997.

Simpson, Lt Col. Mark, "Airpower Lessons from Operation Iraqi Freedom," Briefing, Air Combat Command, Langley Air Force Base, VA, November 25, 2003, pp. 29–30.

Slessor, J. C., *Air Power and Armies*, London: Oxford University Press, 1936.

Smucker, Philip, "A Day-by-Day Account of How Osama bin Laden Eluded the World's Most Powerful Military Machine," *Christian Science Monitor*, March 4, 2002.

Spires, David N., *The XIX Tactical Air Command in the Second World War*, Air Force History and Museums Program, Washington, DC, 2002.

Stanley, Elizabeth, *Evolutionary Technology in the Current Revolution in Military Affairs: The Army Tactical Command and Control System*, Carlisle, PA: U.S. Army War College Strategic Studies Institute, 1998.

Stillion, John, and David T. Orletsky, *Airbase Vulnerability to Conventional Cruise-Missile and Ballistic-Missile Attacks: Technology, Scenarios, and U.S. Air Force Responses*, Santa Monica, CA: RAND Corporation, MR-1028-AF, 1999.

Swortz, Leonard, "Training to Reverse CTC Negative Trends: Getting Fires Back into the Close Fight," *Field Artillery Magazine*, January–February 2002.

Tellis, Ashley J., et al., *Measuring National Power in the Postindustrial Age*, Santa Monica, CA: RAND Corporation, MR-1110-A, 2000.

"Testimony of Steven W. Boutelle, Chief Information Officer and G-6 United States Army Before the Committee on Armed Services Subcommittee on Terrorism, Unconventional Threats, and Capabilities, United States House of Representatives Regarding Department of Defense Information Systems Architecture and Interoperability," 2nd Sess., 108th Cong., February 11, 2004.

Trimble, Stephen, "AH-64 Apache's Deep Strike Role Under Army Review, Keane Says," *Aerospace Daily*, August 6, 2003.

Tyson, Ann Scott, "Elite Air Force Scouts Brave Friendly Fire, Runaway Horses," *The Christian Science Monitor*, March 27, 2002 (accessed on November 6, 2003, at http://www.csmonitor.com/2002/0327/p01s03-usmi.html).

United States Air Forces in Europe, Studies and Analysis Directorate, *The Air War Over Serbia*, Washington, DC: The Air Staff, 2002.

U.S. Army Center for Lessons Learned, "Army Transformation Taking Shape: The Interim Brigade Combat Team," *Newsletter 1-18*, Fort Leavenworth, KS: U.S. Army Center, 2001.

U.S. Army Combat Institute, *Sixty Years of Reorganizing for Combat: A Historical Trend Analysis*, Fort Leavenworth, KS: U.S. Army Combat Studies Institute, 2000.

"U.S. Army Foresees 6,000-plus UAVs for Future Combat Systems," *Jane's International Defense Review*, September 1, 2003.

U.S. Army Public Affairs, "Army Future Combat Systems Passes Major Milestone," Washington, DC: accessed at: http://www4.army.mil/ocpa/read.php?story_id_key=178.

U.S. Army Training and Doctrine Command, *Force XXI Operations*, Fort Monroe, VA: U.S. Army Training and Doctrine Command, August 1994.

U.S. Army Training and Doctrine Command, *Operational and Organizational Concept for the Unit of Employment*, TRADOC Pamphlet 525-3-100, Fort Monroe, VA, November 2001a.

U.S. Army Training and Doctrine Command, *Organizational and Operational Concept Document for the Interim Brigade Combat Team*, Fort Monroe, VA, November 2001b.

U.S. Army Training and Doctrine Command, *Tactical Operational and Organizational Concept for the Maneuver Unit of Action,* TRADOC Pamphlet 525-3-90, Fort Monroe, VA, draft dated November 7, 2001c.

U.S. Army Training and Doctrine Command, *Operational and Organization Plan for the Objective Force Maneuver Unit of Action*, Fort Monroe, VA, July 2002.

U.S. Army Training and Doctrine Command, *The Army Future Force: Decisive 21st Century Landpower*, Fort Monroe, VA, August 26, 2003, p. 2.

U.S. General Accounting Office, *Issues Facing the Future Combat Systems*, Washington, DC: General Accounting Office, 2003a.

U.S. General Accounting Office, *Military Readiness: Lingering Training and Equipment Issues Hamper Air Support of Ground Forces*, Washington, DC: General Accounting Office, 2003b.

van Creveld, Martin, *Technology and War,* New York: Free Press, 1991.

van Creveld, Martin, Kenneth S. Brower, and Steven L. Canby, *Air Power and Maneuver Warfare*, Maxwell Air Force Base, AL: Air University Press, 1994.

Vick, Alan, et al., *Aerospace Operations in Urban Environments: Exploring New Concepts*, Santa Monica, CA: RAND Corporation, MR-1187-AF, 2000.

Vick, Alan, et al., *Aerospace Operations Against Elusive Ground Targets,* Santa Monica, CA: RAND Corporation, MR-1398-AF, 2001.

Vick, Alan, et al., *The Stryker Brigade Combat Team: Rethinking Strategic Responsiveness and Assessing Deployment Options*, Santa Monica, CA: RAND Corporation, MR-1606-AF, 2002.

Warden, John, *The Air Campaign: Planning for Combat*, Washington, DC: Pergamon-Brassey's, 1989.

Wolfowitz, Paul, Deputy Secretary of Defense, "Prepared Statement for the Senate Armed Services Committee Hearing on Military Transformation," Washington, DC: Department of Defense, April 9, 2002.

Woodward, Bob, *Bush at War*, New York: Simon & Schuster, 2002.